The Stars Touch Here:

Part One

Written By: Amilee A. Turner

"Lies, lies, lies, I swear we live in a world of them. I could build me a house of lies I tell you, but no thanks, I'd rather sleep on the street all by myself with the truth."- Amilee Turner

This book is dedicated to my nieces, the light in my life...

Chapter One

It was one of those dreams where I always wanted to wake up, but I knew even if I woke up the dream would still be real. I was in one of those nightmares where other mythological worlds were springing up from every corner. The only problem with that was that this was real. I continued to run, because running was the only option I had left. I was too weak to fight the past I dreaded, and too hurt to accept the unrealistic identity of who I really was now, so why not run? He was an immortal enemy, haunting me in every vision, Dawson was my forbidden love, and my real family was gone. Forever...

I was unaware of how long I had been riding on the bus. All I knew was that it would take me far away for a low price. The bus was nearly empty, with a few people on it, so I placed my long legs up on the row of seats, and laid my head on the window. The windows were covered with that light sheet of frost that vehicles get in the wintertime. I took my hand, and swept along the window to see outside. I knew Kansas was miles away now. Yet, that fact did not keep the knot in my stomach from growing any tighter.

He knew I ran. He knew that I knew my true identity, and what his plans were for us. Worst of all, sooner or later, he would find me and my fate would be similar to, or worse, than my family. I had run from him because he was dangerous and unreal, like some kind of demon. He was a demon. He was dark, cold, and terrifying. I shivered as I saw the bright lights of a distant city come into view. I knew this would be the city where I had to get off, the bus driver's orders. Nothing like being alone in an unfamiliar city, I put my headphones into the side of my duffel bag, and decided to doze off one last time, while sleep would be one of the last things I would soon be getting enough of.

"Excuse me ma'am..."

I heard an unfamiliar, husky voice, and my eyes shot open instantly. The bus driver was peering down at me curiously.

"This is your stop. Please take your belongings. Hope you enjoy your stay in Chicago." He said, following behind me. He sat at the wheel and called out to me as I got off, "Be careful lady, you make it look like you've seen the devil."

"That sounds just about right." I muttered to myself, waving goodbye to him.

I was smack down in the middle of the city, crowded with people, and I was blinded by skyscraper buildings. The breeze was chilly, and my sweatshirt was not doing a great job of keeping me warm. I really did not know where to go at first. There were just too many people wandering about. I had about seven-hundred-dollars with me, saved up from my job as a waitress at a local restaurant back home, but I spent about two-hundred of it from the previous places I had stayed in the past five months, so I knew it would not last me much longer.

"Excuse me sir, do you know the time?" I asked a passerby.

"It is around five this afternoon."

"Thanks."

I decided to get a hotel room so I could figure out what to do. Chicago was vast and full of people shopping and enjoying the tourist sites, especially in the winter, being the holiday season and whatnot. I managed to find myself a taxi, shoving through the enormous crowds of people.

After about a half-hour of searching, and quite a penny to pay, I was dropped off at some small hotel, more towards the outskirts of the city. It was affordable and private, so I figured it was reasonable and a good place to hide.

"How may I help you today?" The short, Hispanic man at the front desk asked.

"I just want a room for the night." I replied,

taking off my hat and sunglasses.

He froze, staring at my swollen eyes. I quickly put the sunglasses back on and opened my wallet, giving him a fifty. He did not say anything regarding the amount of the bill and gave me my key. Dodging his concerned expression, I left the motel office in pursuance of my room. The second I got into the room, I locked the door and turned on the lights. It was small, as was expected. The paint was chipping off the ceiling, and the bed was made with an ugly shade of brown sheets. There was a television set with only five stations, a floor lamp, and a space heater. I dropped my bags, turned on the heater, and went into the bathroom. It was decent for a hotel bathroom, yet small. The tub was clean, but without shower curtains, and the mirror was large. For once, in a long time, I looked at my reflection.

My hair was a massive tangle of long dirty, black curls, my skin, originally a tan color, was now pale and lifeless. I had lost several pounds and my shrunken frame made my tall, hourglass figure look rather anorexic. Worst of all were my eyes. Not the hazel, gray color of them, but the fact that my right eye was surrounded by a huge, plum-colored bruise and my lack of sleep had caused my eyes to swell up. I looked horrible.

"You had a good hit," I whispered to myself. "You killed them, but you will not get me. Not me!" I sniveled, wiping my nose.

Bryce, that was his 'name', had hurt me with some agonizing, powerful force when he found out that I was running away. It was when I had been sitting and waiting at the bus stop. I left according to my plan that day, and I was mortified when I saw him standing a few feet away from me. I remembered his eyes going from their normal turquoise color to scarlet red. The next thing I knew he was chanting some weird language and I was thrown from the bench into the metal trashcan that was nearby. I was sure that I had failed to escape, when some passing men interfered. The men stayed with me until the bus came and threatened to call the police if Bryce did not leave. He was gone instantly, with a smirk, and a glint of evil humor in his eyes. They hadn't a clue that it would take a lot more than the police to stop Bryce.

As painful as the memory was, the bruises hadn't faded a bit. They were still blossoming like new. I took off my clothes and looked at my bruised body. The bruises looked like splattered flowers along my ribs, arms, and legs. Ashamed, I got into the warm bath and closed my eyes, letting the haunting images come back to my mind.

Four Months Ago

"Trish! Not so high! I don't want to go so high!" Nikki, my six-year-old sister, screamed from the swing.

"What did I tell you about fear? You have to take chances and absorb the thrill!" I laughed, giving her another good push on the swing.

"Well I don't want to be scared! Let me off!" She protested.

I picked her up off of the swing and took Henry out of his stroller.

"Look at our little brother Nikki, he's fearless." I swung my one-and-a-half year old brother around in the air, as he giggled. It was July, and like most summers, it was very hot. To make matters worse, it was humid too. No matter, sticky, sweaty clothes or not, I was stuck babysitting my siblings again, while our mother was working fulltime as a local hotel manager. It was the least I could do, to let them have some playtime outside of the house.

"Can we go to our special place tonight Ti Ti?" Nikki asked, as I buckled her up in the car.

"We might if you are good tonight and don't fight with Henry." I replied, turning on the air conditioner to full blast, in our worn-down station wagon.

There was this place that my mom used to take me when I was younger, just outside of Genevah, on a high hill. Many nights, in all seasons, we would walk up there and look at the stars, but low finances forced my mom to seldom take me up there because of bills to be paid and transportation issues. What made this place special though, was the rather large lake near the hill. When one looked down at it, the stars reflections were perfectly scattered across the water. I had started taking Nikki and Henry out there some evenings, just before their bedtime. The place always felt magical to me, especially in the summer when the lightning bugs were out and about. Nikki had given it the name "Fairy Land".

I parked the car just at the end of the hill.

"We can't be long. I have got to get you two into bed, and pick up mom at midnight."

Holding Henry and Nikki's hand, we walked up the hill and sat down on the usual soft patch of grass that had been flattened the many times we had come there before. The smell of the air was that of fresh pine and the soft breeze ran along my skin, giving me goose-bumps. I looked down at Nikki, who was twirling a strand of my hair and gazing up at the stars, and then her big brown eyes turned to mine.

"What are you thinking about so hard?" I said, moving a curl out of her face.

"I want to go there and live up there." She sighed.

"It is quite beautiful. Maybe someday you will." I replied.

"Yeah, mama, you, Henry, and papa will be there with me and we will build a big house out of stars."

I gave her a promising smile, remembering our father's early death just before Henry was born. He had been in a car accident and his head injuries were fatal. My mom had been distant to us children since then, always working and in her own world. I respected this as her time to cope and heal, so it hadn't bothered me much. In an odd way, I sort of enjoyed being the oldest daughter and being able to have a sense of independency and responsibility that many other girls my age did not possess.

Henry and Nikki were the two people that

meant more than anything to me in this world and I promised with all my heart, I'd protect them. I looked down at them both, now asleep, and back up at the stars. At that exact moment a shooting star flew by and I made a wish: a wish for life, a wish for meaning, and a wish for strength. A little too many wishes for one star, but who said there was a limit? Gathering those two small bodies in both arms, I walked down to the car, strapped them in, and drove home.

Around one a.m., when everyone was sound asleep, I lay in bed waiting for the phone to ring. I had just graduated, in May, from high school, and a friend of mine was throwing a summer graduation party just before everyone went off to college. As far as I was concerned, I did not know what I wanted to be yet. I liked art, but I was not sure if I could make a decent living off of it. Not to mention that my mother wouldn't support my dreams of a career in art. She would much rather have me in the medical field. Therefore, college was off the plate for the moment.

My future was uncertain. That did not mean that I wasn't going to party though. Being free from the ties of school was an accomplishment good enough for me. The phone rang. I quickly answered it, as to not wake up my mom.

"I'm outside, come on!" Jenny rushed on the phone.

I hung up. Already dressed in a denim skirt and yellow halter top, I crawled quietly out of bed and out of my window. Sneaking out was the only way of having a social life in my family. My mother gave me absolutely no freedom. Prior to my arrival at the party, the girls I had ridden with, Kelsey and Jenny had gotten a bottle of vodka. They had an older male friend buy it for them. Jenny reached over and offered me a swig of it. I drank occasionally, but I was wiser than most girls I knew. Declining the offered vodka bottle from Jenny, I shook my head, watching her yell obscenities out of the window at random cars passing by. All of this, and we hadn't even made it to the party yet, I thought.

I knew that I would, most likely, be driving them home and walking back to my place, alone and unhappy. That was the basic routine, they got drunk and I, being the loyal friend, took them home. Besides, I did not need my mother having suspicions of a sudden hangover that I might have the next morning. We pulled up to the house. I drew in a deep breath and let it out, as I watched several of the students I had gone to high school with, go into the house. We got out of the car, Jenny and Kelsey grabbing each other's hands while walking towards the door, and I trailed quietly behind them.

It was crowded as usual, with the same people from school, even the long been graduated ones, drinking and smoking around the house, an upbeat music playing on the speakers, and intimate couples here and there on sofas or in corners. The smell of marijuana was strong and I was almost tempted to find the source of it. I lost Jenny and Kelsey instantly amongst the crowd. It made sense that I didn't really have a group of people that I could really call 'good' friends, when I basically had no social life. Parties were rare for me, and if I got the chance, I went with whoever invited me. Tonight, it had been Kelsey and Jenny. I sat down on the couch in the living area and observed my surroundings, sipping on a beer, and trying to ignore the terrible smell of mixed body odors, alcohol, and smoke.

I was getting rather bored, when I noticed an unfamiliar face sitting in a corner with a group of guys from school. The guys didn't seem to be paying much attention to him, although he sat smack down in the middle of them, looking quite uninterested in the conversation taking place. He was different looking than most of the guys that went to my school. His shiny, black hair was tied behind his head, his eyes were an intense dark, ocean blue, and his perfectly chiseled features and lean muscular frame gave him the appearance of a male fashion-model. He was the handsomest guy I'd ever seen, and he was way too good looking to belong in a place like this. Knowing that I had never seen him before, I found myself wondering where he was from and who he was.

I knew he had caught me staring, because his eyes turned and stared into mine. I was entranced by his beauty. I felt like a rabbit sensing the approach of the predator. I was in a deadlock stance and I couldn't even look away from his scrutiny. He smiled at me as if he knew every secret I had ever had. It made me feel uncomfortable and awkward. Embarrassed and suddenly distracted by a sudden police siren in the distance, I quickly got up to go home because I was feeling uneasy in my stomach. Kelsey and Jenny were both sprawled out drunk on a couple of guy's laps. I did not feel like intervening and gave the keys to one of their more sober friends. I told her to take them home, because I wanted to leave.

Outside in the cooler, summer night air, I could actually breathe. My house was about a mile from the party, so I figured I would be okay walking. About a quarter of the way there, I had a feeling someone was following me and I started to panic. How did I always get myself into stupid situations? I started jogging at a steady pace. The night always made everything much more frightening, with all of the shadows from the trees, and dark neighborhood streets. The night always had secrets. Most of them were best kept that way. A black sedan suddenly came out of nowhere and screeched to a stop just next to me. I froze out of fear and prayed to God for my life. It was him, the guy from the party. I saw his amused face as he rolled down the tinted passenger window.

"Did I frighten you? You left the party in quite a hurry... Can I give you a ride?" His voice glided over my skin with such heat that I felt like I was going to faint.

"Um, no thanks, I don't ride with strangers." I replied, the perfect, old-fashioned comeback line.

"Okay, well, I am Dawson and you are?" He

asked, glancing at me, as if he already knew me.

"Trish." My name came out simply before I could even think.

"Well, we are not strangers anymore then, and I'm sure it is perhaps safer to ride with me than to walk home alone. Trust me." He gave me a promising smile.

I wanted to kill myself as I opened his car door and plopped into the leather, black seat inhaling the scent of cigarettes.

"You smoke?" I asked randomly. He smiled at me, mischievously.

"No. This is not my car." He replied, shifting in his seat.

I looked at the several smoked cigarettes in his ash tray and shrugged. He responded, "You do not believe me, Trish?" I cringed at the sound of my name.

"Not at all… So, are you from around here?" I asked, trying to change the subject.

I was uncomfortable, was he a thief or a murderer? My mother would definitely lose her mind if she heard a car pull into the driveway at this time of night. It would be worse if she saw me in a car with a boy. I anxiously held my breath.

“No, I am not. I am from somewhere a bit far from here.”

“You can just pull up—

He pulled up just a few houses down from mine. I looked at him with wide eyes. How did he know about me sneaking out, let alone where I lived? Maybe I had not been paying much attention to where he was driving. Apparently, I was too busy gaping at him.

“You look like a girl that sneaks around a lot,’ he said as if reading my mind, ‘well, good night Trish, and we shall meet again soon enough I am sure.” I froze as he kissed my cheek. It was just the faintest of kisses, but his scent was strong. He smelled just like the air after it rains, fresh and comforting. I got out of the car quickly and nervously.

“Goodnight, Dawson,” I said, letting out the

breath I'd been holding in during the ride.

"Be careful, Trish. Please, be very careful. A young lady like you should never ever walk home alone at this hour." He said, and with that he was gone, speeding down the street and out of sight.

I was suddenly entranced with emotions I had never felt before and a longing for something my eighteen-year-old self was not ready for. I crept back into my room, through my window, locking it, and crawled into my bed. I fell asleep with the last good dream I ever had.

Present

There was a sudden pounding on the hotel room door and I shot up out of the water. My skin had grown prune-like from sitting in the water too long. Taking a quick glance at my bruised body, I slipped a robe on and went to see who it was.

"Uh, this is room service and I brought up some clean towels, and uh, medicine for you uh..."

I opened the door and looked at the nervous,

short woman staring up at me, nervously.

"Thank you. I appreciate it." I replied kindly, taking the items and closing the door.

I went and jumped on the bed and turned on the television to the local news. The space heater had made the room slightly warmer, but I was still chilled. I pulled the scratchy, wool covers up over my body and fumbled with the black star and moon necklace that Dawson had given to me at the funeral. He had told me to never take it off for it was a sacred gift from his father and his fathers before him. It was for safety and guidance, yet right at this moment I knew those things did not apply to me. I was still fumbling with the necklace when I once again fell into the horrid slumber of memories.

I had woken up early that Saturday in July, after the graduation party, and I knew something wasn't right. I smelled cooking coming from the kitchen. No one ever cooked in my house. I got up and threw on some jeans underneath my large t-shirt, and crept out of my room silently. Nikki and Henry were still sleeping in their rooms. I thought that my mother had decided it was time for a change. However, a bigger change than was expected was sitting at the kitchen table as I walked in. First I saw him, a middle-aged man, sitting with his back facing me at the table reading the newspaper. He had brown hair with silver, not white, streaks in it. I was distracted by my mother's voice.

"Trish? Hey dear, come here I want you to meet someone," she exclaimed, so happily that I thought I was in the wrong house for a minute. Either that or I was dreaming.

"Trish, this is Bryce. Bryce, this is my oldest daughter Trish," she introduced, wiping her hands on her apron.

Bryce stood with his back still facing me and I was amazed by how massively tall he was. He had to be seven feet at the least, as he towered over me. He turned to face me and it felt like the world had stopped moving. I froze and I was quite sure my mouth was hanging wide open. His eyes were the first things I saw. They were a brilliant shade of jade and turquoise mixed together. He stared deeply into my eyes and for a moment I forgot I was even there. I shook my head, blinking my eyes. There was a smug grin on his full lips and his teeth glittered, unnaturally white underneath. He looked like a lawyer in appearance, except for the dangerous aura that surrounded him. Never, in my life, had I seen such a fake-looking person. It was like he was wax almost. I must've taken a moment, because I heard my mother cough.

"Oh hello..." was all I could manage as I shook his offered hand.

The second our skin touched, a jolt of fear shot through me. What was wrong with me?

"Well, so much for my first impression," he joked, with a hint of sarcasm.

"Honey, um, I know this may come as a bit of a surprise to you, but Bryce and I have been dating, rather secretly, for the past couple of months and we thought now it would be a good time to introduce you guys to him. I know it may seem a little quick, but I want you to understand that I needed to be happy again." A soft smile of hope played across her small lips.

I was suddenly angry. That would explain her sudden "extended" work shifts and my longer babysitting hours. Since when did she date? I hadn't seen any signs of different behavior and I know I would sense it if she were acting different. Did she not even think of our happiness and the fact that this was just out of the ordinary? Where did he come from? My mother never dated and Bryce did not look like her type at all. I tried to smile, but every time I looked at him and my gaze locked with those scary, multicolored eyes, I couldn't even manage any sort of acceptance at all. She motioned for me to sit down with them, but instead I went to the fridge for some orange juice. All I could think about was dad, and he was gone.

"I made breakfast Trish, so we could all talk about this, I want us to look toward the future as a new family," she explained. That made me bust, and I slammed the carton of orange juice on the counter, making a little spill.

"Maybe I don't want a new family, Mom. I don't think having a stranger interfering in our lives right now, after we just lost Dad, is a good idea. Sorry, Bryce, but I'm not looking for a new father!" I shot at him, as I quickly wiped up the mess I made.

"Trish Rosalyn Noble!" My mother warned, with my full name.

"No, it is fine Diane, really," Bryce insisted, patting her small hands and placing a strand of brown hair behind her ear.

I glared at him. How dare he just come into our home and play father figure! This had to be some kind of a sick joke. The feeling I had about him wasn't a good one, and I knew my intuition never failed me. My mom looked at him as if she was hypnotized. It would normally take my mother months to fall in love with a man, just like it did with my father. Yet, she was crazy about this man she barely knew. Something wasn't right. I knew my mom, and the past two months she'd been depressed and working. I hadn't seen her smile or show any signs of happiness. So how could she have fell in love with Bryce all of the sudden? He turned and smiled up at me, but it did not reach his eyes.

"I didn't come here to be your father Trish, there is a lot more to it than that. I understand your loss and your grief, surely I do, but there is a time when that grief needs to be put away for brighter things." He sounded pathetic.

I rolled my eyes as he kissed my mother's cheek, reassuringly.

"I have to be at work this morning. I'm going to go wash up." I excused myself.

"Don't forget to get the kids from daycare this afternoon Trish." My mom called after me, still gazing at Bryce.

“Have a good day Trish, I’m sure we can discuss this later,” Bryce said, his eyes following me out of the kitchen.

I did not respond. The “we” part didn’t sound like it included anyone else but us. I shivered. That was the day my life changed for the worse. They had me working late that day and I called my mom to let her know how busy we were.

“Well, I’ll be home with the kids until seven, but I need you to rush home right after work. Bryce is going to pick you up okay. It will give you two a chance to make up for this morning’s little mishap. I’m going to come get the car to go to the grocery store. I need to buy some groceries for supper,” she said, worriedly as always.

“Mom, can’t you come get me!” I whined, like a baby.

“Trish come on, how would it look leaving your little brother and sister with a man they barely even know?” She insisted.

"I should be asking you the same question about me," I said.

"Be mature, I'll see you soon. I love you honey," she said.

"I love you too," I replied, glumly.

"We love you Ti Ti!" I heard Nikki's voice and Henry's giggles in the background.

"I love you too, see you tonight!" I hung up the phone.

It never hit me that it would be the last time I would ever hear those words from them again.

Bryce was waiting in a red Lexus as I walked out into the darkening sky after work. I stiffly got into the car and didn't even put on my seat belt. Out of the corner of my eye, I saw his mouth twitch.

"You could make this easier Trish." He said, in a soothing low voice.

"Make what easier? If my mom wants to date you fine, but you are not any part of me or my siblings' life," I said, snobbishly.

The car suddenly jerked to a stop a few blocks from my house. He turned and glared at me. I could make out his bold features and piercing eyes glaring at me in the darkness.

"You don't get it Trish. You are a part of my life and you'll have to get used to it sooner or later," he said, through gritted teeth.

"You're crazy!" I yelled, as I tried to get out of the car.

The door wouldn't budge.

"I knew this would be difficult. I told them you were too young, but they insisted it was you, The Chosen One. It is in your blood, and at first I didn't believe it, but now I do, because you are so darn stubborn. You have the ignorance of a mortal… I have to just have patience with you." He sighed, and reached out to me with his hand.

"You're out of your mind! Somebody help me!" I screamed, as I beat on the window of the car door. He laughed suddenly as the tears started to form in my eyes. I had to get home, something was horribly wrong. He took off just as abruptly as he'd stopped. The second I reached home, I knew that it would only get worse. Home wasn't home anymore. It was a whole scene of firefighters, paramedics, cops, and ambulances. I shrieked when I saw my house, or what was left of it. I jumped out of the car so fast, that I nearly fell. I ran as fast as I could through the crowd of people surrounding the scene. A police officer tried to stop me, but I shoved him as hard as I could.

"Get out of my way!" I ran past him to the ambulance truck, screaming my mother's and siblings' names. There was nothing inside, I turned and nausea swept through me. Three body bags were being carried by paramedics to the ambulance truck. I heard the sound of my body hit the cement. The impact was enough to send my body into shock. Suddenly the world went blurry and then I was falling into a world of blackness. All I remember was asking God for this to all be just a dream, a terrible dream.

"Somebody get her to the hospital. I think she's the oldest daughter. Back up now!"

I heard nothing else.

Chapter Two

I remember awaking to the sound of monitors. I half expected to have been in some sort of accident, but when I didn't see my mom in the hospital room with me, everything rushed back in bits and pieces. An older doctor walked in with a clipboard and was writing multiple things down, with a grim expression on his face.

"I'm going to run a few tests on you and then I'm going to let you leave. There is a man, who claims to be your uncle, waiting for you outside and has insisted you stay with him for a while. You are an adult, but records indicate that he is family. Your mother's brother, I believe. I'm sure it would be best for you two to spend some time together during this difficult time." He said, kindly.

I could barely fathom what he was talking about. Had he just said uncle? I didn't have any uncles, only aunts, on my mom's side. My head was spinning. I was hurt and confused. How in the hell did Bryce get away with something like this, and how could he take my family from me? What did I do to deserve such a thing? I wanted to cry, but the tears wouldn't come. What was going on?!

"Where is my family?" I croaked. My throat was dry. I felt drained of life.

"Funeral arrangements are being made by your uncle, and I will let you two discuss those matters. Everything will be ok," he insisted, patting my shoulder, and motioning for the nurse to come attend to my needs.

I tried to sit up several times before I gave up and had to lie back down. I felt like I was in a nightmare, and I could not wake up. I was sluggish and tired. It was impossible for me to retrace every event of that day and make sense of what had just taken place. It wasn't hitting me that I would never see my mom, sister, or brother ever again. The doctor came and sat down in the chair next to me. Tears began to slide down my cheeks. Why couldn't I have just stayed home with them and this would have never happened?

"Trish, I know this is going to be a difficult time for you, but you cannot let this stop you from living... You are eighteen-years-old now, an adult…" I began to tune myself out of his words of faith. For all I knew my life was already over. The doctor sounded just like an annoying broken record. I had heard this all before when my father died. Death seemed to be everything and nothing to everyone else. I was as helpless as ever and there was no one I could run to. The next few days had been a living nightmare. My mind was completely gone. Without having a chance to protest, Bryce had taken me from the hospital the next day, or more like carried me. I was too weak to fight him and figured that it was my death wish anyway, so why not take the risk?

Bryce had taken me to stay with him at his house, which was an old, rundown farmhouse outside of town. It looked like no one had lived there in years. The grass was near the height of my knees and there was not a sign of life anywhere. The windows were all boarded up and there was nothing, except nature, nearby. The place did not look like it was a place where Bryce would be living, but as far as I was concerned, he was just full of surprises. I wasn't scared of anything anymore. When someone loses everything that they ever had, letting go comes a lot easier, believe me.

Inside the house, it was really quiet and dark. The first day there, he had turned the lights on and I gasped at what I saw. It was as if I walked into an entirely different world. The interior of his house looked like something out of a castle. The whole living room was in a closed stone-walled room, with a lit fire in a massive fireplace. There was, however, only a clock and one large black, leather chair in the room. That was it. The kitchen was bare, except for a large, oak table and two, rather medieval looking, chairs. There were a couple more closed doors on the first floor, and then a flight of stairs that descended upward into darkness. The whole way to his house, neither I nor Bryce had spoken to each other. He seemed to be lost in such intense thinking, that I began to feel gratefully invisible. So when he immediately started talking to me inside his house, I got frightened.

"Your room is upstairs. Follow me," he said, in his dark, deep voice. I had a room? I instantly knew he had planned this all along. I began to cry. He stopped at the foot of the stairs, and then turned to look at me sympathetically.

"What are you going to do to me?" I ask, almost against my will. He ignored my question and then said,

"Don't cry, Trish. Soon everything will begin to make sense and you will understand who you really are." He walked over to me, his huge hands cupping my chin roughly, and forcing me to stare into his fierce eyes. All of the sudden, I felt my legs give out on me, but before I could fall, I was already in his huge arms and being carried off to doom.

That evening I had awakened in a large bed with velvet, silk sheets. I shivered as I noticed that I was not in my jeans and t-shirt anymore, but in a long red gown. I panicked at the thought of him touching me while I was passed out. I looked around the room. There was a lit fireplace in this room as well, the walls of the room were painted black and the carpet was just as black as the walls. There was a dresser with a lamp and not much else. The illusion of the room made me feel as if I were laying in a giant, red bed in complete isolated darkness. I also began to wonder how could the outside of the house look old and vacant, but the inside look like a room from some ancient castle.

Right next to the fireplace was Bryce, the psychopath, sitting shirtless in another leather chair that was red too. I was stunned by how muscularly built he was, and felt uncomfortable being in that rather 'intimate' moment with him. He was staring at me intently, his forehead wrinkled and eyebrows creased together. I wanted badly to go home. Wait, I no longer had a home. I glared at him, with such hatred.

“Are you cold? I can close the windows if you like.” He asked, gesturing to the open window at the far side of the room, where the moonlight was peering through. I glared at him. Why was he acting so nice? What was he, a rapist or a serial killer? Why was this man, who used my mother and killed my family, acting like he cared about me? He was way too old and evil to be acting considerate. I glared at him. All I wanted to do was kill him.

“What do you think you are doing? The police will find you! You are sick and I will not let you get away with this!”

He chuckled and stood up. I winced, feeling a familiar burning sensation on my right side, just beneath my breast. I had it ever since I was little girl. It always burned when I got angry or overwhelmed with emotions. Doctors said it was just a slight, but permanent abdominal pain. They didn’t truly understand it, but they gave me medication to ease the pain. I would've loved to have had it right then and there.

"It hurts, doesn't?" He asked.

"What are you talking about?" I demanded, closing my eyes tightly, as if that would make the pain go away.

"Your scar... Venus's kiss," he sneered.

My eyes widened at him. He had to be clearly insane. What was he talking about? A planet… He came and sat on the edge of the bed and I bunched my legs up, holding on to them tightly, my curly hair falling over my knees. I wished he would just leave me alone. The pain in my side increased the closer he got to me. Trying to stay tough, I held my ground and gritted my teeth. I just wanted to die.

"Trish, it is time you know who you truly are," he began, "You must think that I am a monster, but I'm not you see, you are my birth-marked queen, or you will be. You were born from your mother, Diane, but she is not your real mother or your creator. Diane was struggling to have a child with her husband, and after a series of fertility treatments, she got pregnant with you by surprise. When you were born, the doctors were surprised to find a scar on you that resembled a pair of perfect shaped lips. It caused quite an amount of confusion, because they did not understand how you could have gotten such a phenomenal scar in your mother's womb, so they pronounced it as a birthmark." He explained.

Horrified by what he was alleging, I touched my side in horror. How did he know this? He continued...

"The scar is an eternal kiss implanted on you from you real mother, Venus. It was obvious that you were becoming quite familiar with your heritage as you grew up, because you began to find your place among the stars, I assume. You grew up learning about Earth's philosophies and scientific theories of the planets and the stars; you became infatuated with it.

You do not know that there are worlds and dimensions that humans are unaware of. They do not know of their existence. Trish, you are the daughter of Hejeena, who resides and rules on the planet your kindred call Venus. Your real mother is devoted to vanity and blending into the human society. However, mankind has been protected long enough for their atrocities. It is a shame that she let her kind fall into the human civilization freely. I mean, what was she thinking allowing you to grow up among them? It makes me sick." He spat.

I did not know if I was taking all this information in accurately. I started to really believe I was dreaming, a long terribly, frightening dream. At the same time I felt as if, perhaps, I had gone mad. He looked at me and I could see he was beginning to lose his patience with me. I was not giving him the reaction that he wanted. I do not think I could have if I wanted to.

"At the beginning of time when all the planets

were being created, there was a family of divine rulers created to rule over each of the galaxies, planets, the sun, the moon, and all of these rulers were to watch over the foreign world, Earth, from their kingdoms. There were several complications with Earth, because for one it had no ruler, and the life on it, was different than anything we were used to. It was the world of error that once was the world of perfection. Yet, Father demanded that humans were to be protected at all times, for they were a complicated and doomed species that required guidance, love, and all that other nonsense that their kind could never seem to understand or appreciate. There were several meetings on who would rule earth, but the resolution came down to splitting it between two rulers, Jiles and Xandran.'

'Xandran acted as the outcast of the family, but he just disagreed with Father's ways, and as a result all of the rulers thought it would be good for him to stay in an under-realm where the damned, malicious-minded, and the evil went upon death and were punished, so he could rule over them. Xandran was angry at this, because he did not feel he should be ruling over the evil, because of his exclusion from the family, and he sought revenge. He decided to have an affair with a mortal woman and this brought chaos among the rulers, who kept their identity secret by oath. He had five half-immortal children with her, and they were sentenced to stay with him, while their mother was killed. This upset Jiles, who ruled the sky, and he protested against the killing of an innocent human being as a result of a ruler's misbehavior under his watchful eye. He ordered another meeting and there was to be a compromise made if Xandran was going to accept his position in the under realm.'

'Venus had the most beautiful daughters of any of the planetary rulers, and Xandran wanted a queen, but more than that, he wanted the most beautiful queens for all of his sons. Venus objected heatedly on this option. She made it clear that if one of her daughters were to be given to Xandran, she would start a disastrous war. Out of fear, she went to Mavus, who rules the moon, and secretly negotiated with him. He regretfully agreed that he would command his sons to be protectors of her daughters. The only problem with that little negotiation was that she was still outvoted by the other rulers. So Venus, with all her dignity and arrogance, started an endless war with Xandran. The other rulers isolated themselves from the feud, every now and then, would they offer some alternative for peace. Each failed every time, so they left it up to Venus and Xandran to solve their dispute on their own. Venus's daughters are still safe with her right now, however, Xandran told her that there was only one daughter of hers that he was truly interested in, she would be the most powerful, and was yet to be born. He would do whatever it took to capture her for his son next in line for his throne." He smiled at me.

I swore that my heart had stopped beating.

"Centuries went by and the war settled down, that is, until you were born. Venus apparently lost her mind. She knew you were the 'Chosen One' the second you were born. She kissed you with her lips, planting that eternal sacred kiss. That pain you always feel is the radiation of her power, protecting you, but that will be gone soon. Then with all the power she had, she gave you to the average woman here on earth. She gave her destined immortal child to a mortal woman in a town that no mind would ever wander across. It was quite intelligent, to be honest, almost no one knew you existed. We all knew the 'Chosen One' was going to be born, but we did not know she was the one to conceive you. However, your mother made a big mistake. She was quite witty because she picked a good, safe location for you to grow up healthy and pure. It took me quite some time to find you, but Venus did not realize that Xandran and Jiles rule this planet, and that any unfamiliar scent upon this soil is known to them and their companions. I had to, of course, wait until you were of age to come and find you, because you stop aging in a few years, when you are twenty-one, and it would be quite awkward for you to have seen your family go on without you. Little Nikki living on to be eighty, while you are still twenty-one, not a pleasant vision is it? So, I had to make this less painful for you and get rid of them before time ran out." He scooted closer to me.

I just stared at the fire, unable to understand what this information meant. What in the heck was going on?!

"It is not half as bad as it sounds. You will find out soon enough, for my home is your new home,' he softly smiled, then stared, deeply into my eyes, 'I am Xandran's son and next in line to be ruler. Time is running out, our days are numbered, and you are going to be my queen. It would please me if you would quit staring at me with such horror and disgust for I have done nothing wrong, and have saved you from any further grievances." He leaned forward to grab my hand. I jumped off the bed and backed up against the wall. Fear did not even begin to cover it. I began to tremble. He stood up and suddenly he was right in front of me. How did he do that?!

"Now there are no more hidden secrets. You know who you are and who I am. This can be easy or difficult, your choice." I tensed with fear, as he bent over and kissed me. The second our lips met I felt a blast of heat throughout my body and then darkness. I was beginning to get used to passing out.

The next day had been the funeral. I was still shocked by everything I had found out the night before, and even worse having trouble believing it. Nothing seemed to make sense anymore and I felt weak and vulnerable to everything. I felt like I was living in a world that was unreal, except it was real, and I regretted not ever having the faith to believe in all of the possibilities that life could bring forth. Who knew reality and unreality went hand in hand? Not me. Bryce had made me wear a black gown. He made me bathe, comb my hair, and put on makeup as well. I felt unable to make any choices on my own, not that it was an unfamiliar feeling anyway. I half expected to hear Nikki yelling at me to wake up, but I knew that it would never happen again. I suddenly began to feel that this life was not my own. I did not have a life of my own anymore. I was a zombie, cold, empty, and lifeless.

The funeral was at the local cemetery. I had gone to each of their caskets for a final goodbye, withstanding the humid, sticky heat of July. Henry was peaceful looking in a blue, tiny suit. I held my breath as tears slid down my cheeks. Of course they were only sleeping, I tried to reassure myself. It was evident that they had died from the smoke and were rescued before the flames had reached their bodies. I bent over and gave him a kiss on his clammy, lifeless skin and moved on to Nikki. Closing my eyes, I tried to withhold the vomit forming in my throat. I paused at Nikki in terror. Her eyes were close, but her face was twisted in pain, like she had been hit hard. Acting on reflex, without thinking, I lifted her little cold, dead body into my arms and hugged her tightly, and it felt comforting. I heard gasps and whispers from around me, but no one spoke to me.

I felt the harsh stares of family members and close friends of my mother and father behind me, but I did not care. They were my mother's family and never ever associated themselves with me. The main reason being, that they did not like the fact that I looked so different from my mother and siblings. I was the 'alien' daughter. That was an off-hand joke one of my aunts used to make about me because I did not resemble my mother or my father. I had actually begun to wonder if I was adopted growing up, because I did look different compared to them. Now I knew why. Tears fell down my cheeks, as I laid Nikki back down and moved on. I was the last survivor, and I had let them all down. They probably wished I was dead too, well so did I. There was no way I could continue to do this, and how could I expose Bryce? They wouldn't believe me anyway if I told them that the devil was trying to take me away.

"I will meet you in the stars Nikki. I promise." I whispered, to myself.

I went to see my mother, lastly, looking at the woman who raised me, and I knew this was the woman that I would always and only consider to be my mother. I was now in a situation where I did not have a clue what to do. I wanted to scream at everyone there and tell them what Bryce had done, but something didn't feel right about that. Everything was wrong. My mother's family all sat there amongst themselves pretending to be engaged in their own conversations, but I could feel them glancing at me cautiously, as if expecting me to go ballistic. They all thought that I had killed them. They thought that I had burned the house down! All of the sudden I did not want to be there anymore, not at this funeral. I needed to leave, desperately.

Backing away from the caskets, I looked around for Bryce and he was nowhere to be seen. I looked at all of the mourners and the priest speaking much about nothing. As he rambled on about the afterlife and how to cope with grief, I spotted Dawson leaning against a tree, staring at me with a bothered expression. He was dressed from head to toe in black, and looked graceful and comforting. Not giving a second to question what he was even doing there, I ran to him as fast as I could. He caught me swiftly as I landed into the hardness of his chest. I looked up into his familiar ocean, blue eyes and the tears began to pour again. The beginnings of a breakdown were taking place. Why did he feel like home?

"Listen to me, Trish," he urged, stroking my cheek with soft fingers. He pushed me gently away and leaned toward me.

I looked at him through tears. He pulled a black star and moon medallion out of his jeans pocket and placed it around my neck. I looked at him confused, blinking through tears. He came to my family's funeral to give me this?

"Never take this off. As long as you have this, I will know where to find you and you will be safe. It is for protection, passed down from my father and his father's before him. Now listen, I need you to leave. Just run. I don't care where you go, but catch the bus and leave town tonight. I will not ever be far, but you must understand that he has companions. They are demons that follow his every command and they know who I am. I will explain more to you later when you are a safe distance from him and he will not know what we are discussing. He is near enough though, and can hear and sense things that are suspicious. I know you know who you are now, but there is so much more I need to tell you. I believe in you. Leave Trish, and I will find you soon." He said, loosening his grip on me.

So it was true what Bryce had told me! I pulled him back roughly into an embrace. Dawson knew this was going to happen all along! He had probably been watching me for a while. My mind flashed back to the party and how he knew exactly where I lived and my sneaking out regimen. Surprisingly, I wasn't even angry at him, but more so relieved. This just meant that I wasn't going crazy, and Dawson may even have been here to help me.

"Do not leave me Dawson." I begged, digging my nails into his shirt. "I need you! I need you so much! You don't understand. I don't know where to go? I do not know who I really am anymore." I cried.

A hard line formed on his mouth and then he hugged me. He pulled my face close to his and our lips almost touched. I yearned to kiss him, to pour my pain into his lips and have him take it all away. All I wanted was, for someone to take away all of this frustration, confusion, and suffering right off of my back. I couldn't handle this on my own and I had no one.

"I know. Look, I promise I will be near. Trust me, we will meet again soon and there will be time for explanations, but not now." He kissed me, for just a second, and before I could say another word, he was gone. It was as if he had disappeared into the air itself.

I felt something bite my arm. Acting on instinct, I smacked the mosquito and saw its smashed, bloody body on my arm. I groaned and as I did, the medallion lit up. My jaw dropped wide open. I definitely knew that I was not dreaming anymore.

Chapter Three

I woke up startled and was aware that I was really cold. Then it hit me that I was in a hotel room with a terrible furnace, and that I had no home. I thought of Dawson. It had been five months since I'd seen him in July, and I had stopped believing I would ever see him again. I looked at the curtains and saw the sunlight peeking through. I would be checking out the hotel for another night. My money would be gone before I knew it. My anxiety was bad, real bad. I did not know where I was going. At this point, my life was unclear.

As I lay there, I stared at the ceiling and remembered my last encounter with Bryce. I had tried to trick him into believing that I needed to go into town to purchase a couple of things. He had said not a word when I made my request, sitting alone in his solitariness downstairs. His eyes never looked at me and he continued to stare at the bare fireplace, in deep thought. He held out to me the keys to his car too easily. I knew he was not foolish, but I wasted no time leaving. Why did he not come after me? I parked his car in the downtown parking garage and half jogged, in and out of several stores, grabbing items for my escape. I stopped by my bank to withdraw my savings from work. It appeared that everything would go according to my plan, and I would be getting on the bus in the early evening. I left his car where it was and never returned to the house. He had known I would not be coming back too.

When I had been waiting at the bus stop, I felt a strange presence near me. No faster than a second, Bryce had me by my arm, making me drop my luggage. I managed to scream when he threw me to the ground roughly, his face was scorching with rage. To my luck, there had been a group of men walking out of a diner, who witnessed what had taken place. They told him to get lost or they would call the police. He had slightly smiled and apologized as he let me go. The men stayed there with me until the bus came. I did not speak to them, though I wanted to, the fear apparent on my face. The men, seeing this, did not ask any questions about who he was, but rather just were concerned that I got on the bus safely. Even though Bryce had left, I knew he was there watching me leave, hidden, against his will. It did not matter that I was leaving, because I knew he would find me, and just the look in his eyes showed that he knew it too.

I decided that I could not lay in bed another minute. I was tired of remembering all of those memories. I needed to breathe. I needed to know I was still living in a real world. I got up and changed into a clean pair of jeans and an oversized sweater. I was still upset that the bruises had stayed for months, when they should have been gone by now. It was like Bryce had planted permanent bruises all over me. I let my, now, longer hair fall down over my shoulders with the scent of shampoo bouncing off of my curls. I threw all of my stuff into the closet and bundled up in my jacket, to go outside. The icy breeze of the mid-afternoon felt nice against my warm skin. I walked into the lobby to pay for another night.

"Do not worry about it, miss, it is on me. You can stay here for as long as you need. You look like you have had a rough trip." The manager said, sympathetically.

"Thanks... I really appreciate you kindness." I said, tears welling up in my eyes, as I put my wallet back in my pocket, and walked back out into the cold winter air. It had been forever since I'd felt anyone have even the littlest bit of sympathy towards me. I had traveled everywhere and somehow managed to end up in Chicago, closer to home then I had expected. Maybe I should have stayed in Virginia.

I caught a bus into the city, and actually thought about finding a job and starting life over. I had not heard from Dawson, and Bryce apparently had lost interest in finding me. It seemed almost as if the nightmare was coming to an end… almost. I looked around at all of the shops and the after New Year's sale displays. It did not really feel like a new year at all. Time seemed to have flown by fast. I walked past groups of people, several with their families and loved ones, and felt the urge to scream. I wanted to be heard and wanted someone to help me for a change. I was tired of being on my own and not having a clue about who I was. The sun was already starting to set when I walked into a coffee shop and sunk into a booth. The waitress, a tall slim girl with spiky, pink and green hair, came to take my order.

"What can I get you?" She asked, nibbling on her lip ring.

"Something warm and energizing, like coffee." I replied, gloomily.

She smiled at me and nodded as she walked back to the counter. I closed my eyes tightly, trying to figure out what I was going to do with my life. A few minutes later, someone tapped on my shoulder. I looked up. It was the waitress with my coffee.

"Thanks." I said, taking the warm cup with

both hands.

"Eh, don't mention it. By the way, I'm Gabriella." She curtseyed, for whatever reason, put down the cream and sugar, and went back to work.

Gabriella, I thought to myself as I mixed the cream and sugar into my coffee. I did not know what it was, but I ended up staying there until Gabriella got off work. I just wanted someone to talk to, and she just so happened to be my person of choice. After three cups of coffee, a piece of toast with grape jelly, and a dull period of tearing up straw papers, while I stared at Gabriella all night, she was finally off. As she was putting on her coat and walking out the door, I ran up to her.

"Uh Gabriella, I...Would it be possible to talk or something?"

She laughed. I looked at her, questionably. Was that funny?

"You must need to. You've been staring at me all night! Sounds cool, you want to ride?" She asked, not taking any time to even wonder why I, a complete stranger, had waited for her.

I looked at her, confused, and just nodded my head.

We walked out into the cold, night air and her hair glimmered in the lights of the city. She kept walking and I just followed her. She popped a cigarette into her mouth and turned her bright, green eyes to me.

"So where are you from?" She said, exhaling the smoke into my face.

"I'm from Kansas," I said, solemnly.

"You don't look like you are from there." She said, offering me a cigarette, which I declined.

"Well unfortunately, I've learned that I was not originally from there."

She nodded and kept walking in silence, until we reached a small parking lot.

"This is it. Hop in." She said, unlocking the doors of a blue minivan.

I got in without a word. She did not even ask

where I was heading to, she just drove. We ended up parking in front of a worn down apartment complex just on the outskirts of the city. I followed her into the building. It smelled of urine and sour musk. The wallpaper was peeling with graffiti on it, and the floors creaked. We walked up a couple flights of stairs and she unlocked the door to her apartment. I held my breath as we went in. Gabriella's apartment was surprisingly, nice. It smelled like incense, and the floor was nicely polished wood. She threw her coat and bag onto her brown couch. I just stood there, while she draped her black curtains over the windows and locked the door.

"You can relax, you know. Here, let me take your jacket." She said, reaching for my arm to help me take it off. I flinched as she touched a bruised spot.

"Sorry, did I hurt you?"

"No, I'm okay. I like your place." It was nice and homey. I actually felt, in a sense, normal again.

"This is a piece of shit. I have to give the

landlord a blow job every month, so he can give me a discount on rent," she replied, walking into the small kitchen, and pulling a bottle of water out of the fridge. I looked at her in astonishment. Had she really just said that?

She laughed and motioned me to sit at the small table in the room. I glanced around at the pictures of rock legends she had on her walls as I took my seat. She lit another cigarette, the fumes filling the room.

"So tell me about you and what's going on." She said, staring at me curiously, with her amazing green eyes, as she ashed her cigarette in the ash tray.

"I do not know myself. I mean, I just lost my family in a fire recently, and then had found out that they weren't really my family to begin with, and that my biological mother is actually the ruler of the planet Venus. You do know astronomy right? Well I could not believe it when this man, who I have been running away from, told me—." I caught my breath, looking down at my hands, fidgeting…great… I couldn't bring myself to finish.

"I see." She said quietly, inhaling her cigarette, and blowing the smoke up in the air.

"I know it sounds really ridiculous, but I have nowhere to go, I am running from this demon who is trying to marry me, and the only person that can help me is not anywhere to be found. For the past several months, I've done nothing but travel, sleep in parks or in hotels, and eat nothing but junk. I do not know where I am going and I cannot understand what is happening to me!" I began to softly whimper.

"Wow this is tough stuff… I do not know how to respond to this. Um... are you mentally ill? Just asking," she asked concerned, raising her eyebrows.

I shrugged, rolling my eyes. This was hopeless.

"Probably… I should get out of here, this is stupid of me. I apologize." I said, getting up.

"Hey now, I suggest you stay. The crazies are out there at this time of night. I want to help. You know I do not think you are insane. A little weird, maybe took a few strikes to the head, but I like you. You should stay here for a bit. I can get lonely sometimes too. That is, unless you have somewhere else to be." She took her ashtray and dumped it into the sink.

She motioned me to follow her, and she led me into her bedroom. It was exotic. Her bed was covered with dark green covers, and the walls were painted a rich brown. She had a dresser set and plants positioned around the room neatly. There were picture frames on her vanity table, and a large poster of Kurt Cobain on her wall. A black cat lay, curled up, sleeping on her bed.

"I love your room." I said, admiringly.

"Me too," she said, undressing.

Unable to help myself, my eyes drifted over her body. I noticed she was really pale and skinny, as she comfortably slipped on a clean pair of underwear. She threw on a black tank top to match her underwear and lit yet another cigarette. I wondered how long it would take before those things would eventually kill her. She threw a t-shirt at me and some shorts. I started to undress, self-consciously.

“Holy shit!” she gasped, making me drop the shirt, “What the hell happened to you?”

“I had an encounter with the demon that I was telling you about. I was trying to get away from him.” I answered, anticipating her response.

The truth seemed all I could say. I looked at my reflection in the mirror. The bruise under my eye seemed to have faded away, but the others were still there.

“Well damn, looks like the devil to me. Let me put some ointment on them,” she said, grabbing a jar off of her dresser, and forcing me to sit down.

Who puts ointment on bruises? I thought to myself.

“Look, I know that I do not really know you and all, but you seem to be in some serious stuff, and I am not all into the fantasy shit, but this is crazy. If there is a man or demon as you say beating you, then honey you need to report him or find a preacher." She looked at me sternly.

I felt the smoke of her cigarette caressing my back as she applied the ointment.

"You know, if I were you, I probably would not believe my story either, so it is okay." I responded.

I looked at the pictures she had on her desk. They looked like family. Then I saw one of a little girl with the exact same green eyes as Gabriella. Gabriella was holding her, they had ice cream cones and they were laughing in the picture. Gabriella noticed me looking and she put the jar down and spoke.

"That was my daughter Zeo. She died a few years ago…" she said, softly.

"I am so sorry Gabriella. How old was she?" My heart sank.

"She would be five this year in November. We had the same birthday." She smiled softly, reminiscing.

"How did she die?" I asked, out of curiosity,

wishing I hadn't.

"She was shot during a drive-by, while she was playing in a nearby park," Gabriella said quietly, "It was just her being in the wrong place that should have been a safe place, at the wrong time and I not getting there fast enough." I could hear the regret in her voice.

I could tell that she blamed herself for her daughter's death and I felt tears well up into my eyes. What was it with death? It just seemed like the easy way to ruin a person's whole point of existence. Gabriella got up and put another cigarette in her mouth, going to her closet to get a bottle of brandy from her stash. I wiped my eyes, quickly, not wanting to make a scene.

"Drink?" she asked.

What did I care for?

"Yes, that would be nice."

"I normally go to this club on weekends, free alcohol, hot guys, and whatnot. You should go with me. How old are you?" She asked, pouring me a small glass and giving it to me.

It took me a minute to remember as I took a sip of the brandy. It burned, nicely in my stomach. My nineteenth birthday had already passed in late September. I had been consumed with running away and hiding, that I hadn't even noticed the passing time. The New Year was constantly hitting me in the head, trying to get me to wake up.

"I'm nineteen..." I replied.

"A little younger than me, I'm twenty-three. But I am sure I could get you a drink without problems." She said, getting under the covers.

"You can sleep with me or you can crash on the couch, it does not matter. We will go tomorrow. Get our minds off this shitty life." She gulped down another drink from the bottle, and closed her eyes.

I got up and walked to her bed, the cat hissed at me as I got in under the covers with her. It was warm. I did not want to be alone and Gabriella was here. She turned off the lights and I felt her put her arms around me. I stayed awake and stared in the darkness, listening to Gabriella's heartbeat. What a day it had been! Never had I expected to run into someone like her. I felt like fate bonded me with her. She was intriguing and I was attracted to her painful, but understanding personality. I was so tired. Slowly, I fell into a deep slumber and my last thoughts were of Dawson, and he was yelling at me, telling me that something was coming.

Chapter Four

Gabriella had dressed me up for the club that evening. I hardly recognized myself. I was nowhere near her size, so I wore some ripped jeans that she had from her former pregnancy. I was wearing a skull belly shirt with an oversized, leather jacket. My curly hair was now in a sleek, straight look, and I was finding it hard to stand the false eyelashes and bright red lipstick that Gabriella had applied on me. I was never really a rock or punk kind of person and I never wore my hair straight. I glanced over my shoulder at the length of my hair, almost touching the top of my bottom. My mother would have killed me if she saw me like this.

When we arrived, Gabriella held my hand as we walked up past the long line of people waiting to get into the club. I felt the icy stares of men and women, who had been waiting a long time in the freezing weather, as Gabriella dragged me up to the doorman.

“Hey, Doug, what’s up? This is my girl here,

we want in," she said, blowing smoke into his face and clicking the spikes on her leather boots together, signaling that she was being impatient.

He looked her up and down suggestively, and I wanted to slap him. Pervert.

"Go on in, but introduce me to your lovely friend." He smirked, bulking up his muscles in his too tight security shirt. I remained silent. She looked at me and then put her cigarette out.

"No honey, this is my date. That would be against the rules." She grabbed my hand as we entered into the neon lit club.

The sudden blast of heat and sweat felt good apart from the frozen air outside. The place was crowded. People were dancing on tables, since the dance floor was already packed. We could barely move. There was some band, I did not even know, playing on stage. Booths were filled with people literally having sex with each other. If my mother was still alive, she would have killed me for being here. All

I could think about was my mother and how disappointed she would have been with me, with everything that I was becoming. I suddenly had a huge headache. The neon lights and deafening music were making me dizzy.

"Come on, Trish, let's get a drink. You need to sit down." Gabriella said, pulling me through the mass of people, to the bar, where we were lucky enough to find two, unoccupied stools.

"Hey Jeff, we will have two Bloody Mary Martinis! Spice it up for my girl here, she's new." Gabriella screamed over the music.

I shot her a questioning look. She didn't look like a martini drinker, and I certainly wasn't, but oh well. The cute bartender winked at me and got to work on our drinks. He was back in an instant with our martinis and two shots of what appeared to be tequila.

"Here to you sexy ladies. The tequila's on me," he said, winking at me.

I reached for mine and he grabbed my hand.

"What's your name, babe? You want to get to business later, maybe go to my crib?" He leaned toward me. His breath reeked of alcohol and vomit.

"Fuck off," I said, jerking my hand free.

He walked away, insulted. I took my shot and looked away. Gabriella was too busy watching the singer of the band. He happened to be staring at her too. He was tall and skinny, just like her, with sandy, blonde hair drenched in sweat.

"That asshole was Zeo's dad!" Gabriella screamed in my ear.

I could tell she was getting drunk, because she kept yelling obscenities over the music at nothing in particular and downing shot after shot. She had also taken a few drinks of vodka before we came. I did not want myself to get that out of control. She had forgotten my existence.

The band finished their song and another band took their place. Gabriella yelled that she would be right back and walked off without a word, toward Zeo's dad. I looked around, and suddenly felt like I was going to puke. I did not expect Gabriella to just take off like that. Everything was going too fast, and I was certain I was going to faint. I made my way out the back door and into the alleyway. The cold air felt good against my heated skin, as I bent over against the side of the building gasping. The smell of garbage was foul. There were a couple of men nearby smoking cigarettes and eyeing me intensely. I had an idea of what they were thinking and I was beginning to feel uneasy in my stomach. I was about to make my way back into the club, when one of the men, came and stood in my path. He was a large man and was very drunk. Not a good combination.

"Hey baby, come over here.... we want to talk," he laughed, choking on whatever liquid was in his mouth.

"Leave me alone." I said, icily.

"Oh hey now, watch the attitude honey, we just

want to have some fun," he said, moving closer.

The other man slowly began to move forward as well, laughing like a hyena.

"Be nice and this will be over in a quickie." The fatter one said.

I looked at them and it did not seem to hit me that maybe I should scream or take off running. I was not even scared. My head hurt and I was tired. I collapsed on the ground and started to cry. What did I care if they raped me? Who was I anyway? Not anyone important apparently. I felt weak and alone, and helpless. A huge part of me was just tired of living and running away. Death was welcoming at this point.

They paused for a second, glancing at each other, unsure. Then they started arguing on what to do. The next thing I knew, both men let out a piercing scream and then there was a deadly silence in the night. I looked up from crying to see their lifeless bodies sprawled out across the ground. I stood up, suddenly scared. The next thing I knew, I was on some man's shoulder, he was running really fast and effortlessly with me on him. I could not make out who he was, because he was dressed from head to toe in black. He suddenly dropped me on the ground and grabbed my throat tightly. I stared at him, trying to breathe. I saw the faintest glimmer of yellow eyes from behind his mask. I was too weak to fight him. My body began to go numb. Tears dripped down my cheeks, as I felt my last breath come to an end.

I woke up in darkness. Wherever I was, it was freezing. I was lying on a cover and there was a pillow under my head, but I was still on the ground. I groaned, as I lifted my aching body up off the ground. I reached for something to grasp onto and there was nothing. I searched and searched until I bumped into a rocky wall. It was smooth and wet under my fingers. Was I in a cave? I sat back down suddenly terrified. Where was Gabriella? Where was I?

"Hello?" I croaked, and my voice was hoarse.

There was silence. Then a door, from nowhere,

opened and a dark, cloaked man stood in the doorway.

"Come. My father and brothers wait for you," he said, quietly.

I got up and walked towards him unsure and lost my balance. I stumbled into a candlelit hallway that resembled a castle. The hallway was long, and there were large wooden bolted doors lining both sides. At the end of the hallway was a large door with a crescent shape symbol engraved on it. The cloaked man stood there waiting. I looked at him, and I could barely make out his yellow eyes, peering at me from underneath his hood.

"Come. You must make haste." He gestured me to the door.

I stood in front of it. He knocked twice, then three times more, then twice again. Then he turned around and walked back the way we came. I stood there frozen, not knowing what to do.

"Where are you going?" I called after him.

He did not stop to respond, he just kept

walking. The door in front of me suddenly opened. I did the only thing I could do, I walked in.

There, before me, was a large room. It looked like a huge stone castle room, except a long rectangular portion of the stone flooring, in the middle of the room, had been dug out and was filled with sparkling blue water. There was a little bridge that went over it, that led to a small platform with a stairway on it, leading up to yet another platform, with a huge, crystal orb globe at the top. I did not know what to do. It all seemed like a dream. Never having seen anything like this in my life, I took a step onto the bridge and started walking towards the stairs. As I glanced down at the water, I could see stars and planets all over. I began to feel a yearning in my heart, and all I wanted to do was jump into that water and drown.

The bridge began to tremble and change, making me stumble to catch my balance. I hurried across it and up the stair to the first platform. I stopped to catch my breath and look back at the water to see why the bridge had trembled. There was nothing behind me, but a glass wall. I looked at my reflection in horror. My mascara and eye makeup was smeared across my face, with dry, black trails where the tears had dried. The false eyelashes had fallen off a long time ago. My hair was a massive tangle of straightness and curls. I looked like a monster. This was by far the worst nightmare I had ever had. That was when I heard heated debate from nowhere.

"Father, she is the Chosen One. She is Venus's daughter. You cannot simply give this girl to Xandran and Bryce. As much as you are in agony with Venus, you have to understand that Trish holds remarkable power. This could only lead to destruction. She needs to be with her kind. We gave a vow to protect her. We cannot fail our word. I know I have not been this sincere or protective about one of Hejeena's daughters in centuries, but now that Trish has been found, I know she is the one Xandran seeks for his son. She is the Chosen One. You have to see her for yourself. If Bryce gets to her, he will hold enough power to attempt to take over many galaxies. That could lead to a terrible war between our neighbors and our own kind. So far Bryce is on the search, but he is taking his time and enjoying it," an unknown, yet familiar voice said.

"Son, I am aware of our vow, but for all I know we cannot be sure she is who we think she may be. Venus has been losing her head since the child disappeared and her family was killed. The girl's whereabouts are unknown. You have had difficult time finding her, which was your assigned job. I cannot trust this choice. For crying out loud, one of your brothers found her in an alley way of a nightclub in Chicago!" A much older man, fumed.

"Father, I understand, but you have to believe me. I had a strong feeling of her and where her presence was located and I cannot explain why I know she is the one!" The other voice spoke, defiantly.

I heard a tired sigh and then, "Son, all I am saying is that if I bring the wrong girl to Hejeena, that woman will have my head. This could be Xandran's form of trickery. It would be a great relief for us if he is seeking the wrong girl, and believing she is the Chosen One. Once Bryce crowns a queen, she is his queen forever. The only way for them to be separated after oath is by death. If he crowns the wrong girl, then we have nothing to worry about anymore. I will need to speak to and see the girl for myself. It is unnerving that you seem to have more than necessary emotions for her...."

I crept my way up the stairs toward the large crystal ball. When I reached the top, I was completely confused. There was nothing there, but a crystal ball. I had expected to run right in on the two people speaking, one sounded so familiar. I leaned toward the ball, entranced by its light, and sparks began to fly off of it. Unable to help myself, I lightly placed my hand on it and the floor started spinning. I fought to keep my balance, as the crystal ball slowly sank down into the floor and in its place, up came, a wooden door. Getting up on my feet again and regaining my senses, I tried the little gold knob. It turned and the door opened. Without even thinking, for everything was bizarre to me now, I walked inside.

I stopped dead in my tracks as I entered a small room. At a desk, stood an old man that looked just like the typical version of Merlin, from the stories I used to read when I was younger. He was in a long, yellow robe and his beard went all the way down to his toes. There were books and celestial objects everywhere. It looked like a librarian pit. His yellow eyes stared at me in surprise and he crinkled his nose in disgust, all of his wrinkles following suit. Then I saw him. Dawson in all of his charm, standing there staring at me with a look of concern. His ocean blue eyes were still the same and his longer black hair reached just past his shoulders. He was wearing a black robe. I felt the heat rush to my cheeks as they both looked at me, up and down.

"This is Trish, the Chosen One?! The powerful

daughter! You have got to be kidding me!" The older man mocked, glaring at me with those intense yellow eyes, and then looking at Dawson in disbelief.

"Trish, you are up?" He stated, rather than asked, ignoring the other man's sarcasm.

I was speechless.

"This is my father, Mavus." He stared at me, urgently, expecting me to say something.

"Hello, I am Trish," was all I could manage.

I stared at Dawson and fought the urge to run to him. Why was I acting like an idiot? All I wanted to do was hold Dawson. He shifted his weight from one foot to the other foot, then stepped a little closer to me, as if he sensed what I was feeling too, and wanted to grant me my wish. It had been too long since I had last seen him, that I could not bring my eyes away from him.

"So you are this Chosen One eh? The one Hejeena spent centuries fretting about." Mavus scoffed, bringing my attention back to him.

He looked at me disapprovingly.

"So tell me, Trish, do you know who you are?" He asked.

"No." I replied, fighting back tears.

He glanced at Dawson, disturbed.

"Son, I want you to take her to 'The Falls' to get cleaned up. Both of you need to meet me afterwards so we can discuss this matter," he said, giving me another hard look, and then he vanished. How did they do that? Just disappear, in the blink of an eye.

Dawson was suddenly in front of me. He kissed me unexpectedly. I devoured his kiss. It was like eating a forbidden fruit. I did not want to stop. He held me fiercely and I felt his nails dig into my back. When we broke free, his blue eyes were the darkest blue, almost black. It made my eyes go wide. He smiled at me.

"Trish, I told you I would come back for you, why did you doubt me?" He asked, stroking my face.

"You blocked your mind from believing in me. I was unable to get through to you. That is why it took so long to find you. After several attempts, I was able to use telepathy, and communicate with you in your sleep. That is why I was capable to seek where you were located. I sensed your presence." He cupped my cheeks and laughed, relieved that I was okay.

"Dawson... I am so lost..." I said, quietly, more tears sliding down my face.

I shook my head and sobbed into his chest. He held me, just because it was right and because there was nowhere in the universe we would have rather been, then in each other's arms at that moment in time. After we broke free from our embrace, Dawson took my hand and guided me out of his father's chamber and back to the room with the crystal ball. He, as I had earlier, placed his hands on the ball confidently, and the room transformed into another part of the cave, with a stairway leading into darkness. He took my hand into his strong grip, and we began to descend down the stairs.

"Dawson, what is 'The Falls'? What does he

mean I need to cleanup?" I asked, confused.

Dawson let out a huge laugh.

"Trish, look at you, seriously. I mean what is with the leather jacket and scary makeup. I had almost thought I had the wrong girl captured." His voice echoed.

He pulled me closer to him, as we descended down the steps. I blushed, with embarrassment. Apparently, I had forgotten that Gabriella had given me a makeover for that evening. I was about to explain, when I began to hear drops of water falling. We halted to a stop.

"Shh… Now this place is sacred. My father is feeling sympathy towards you, because no one, except my family of course, is allowed to cleanse in here." Dawson whispered, in the darkness.

"Dawson, I do not care about cleansing anything. I need to know answers. I need to know what I am supposed to do. Why is Bryce after me? What is the Chosen One? I've never felt like such a nutcase in my life!" I started babbling away with all of my inner thoughts that I had held within. Dawson grabbed me by the shoulders and gave me a rough shake. I still could not snap out of it.

"Trish, I need you to focus. My father and I will explain everything to you at the meeting. For now I need you to be strong, okay?" He grabbed my hand once more, and pushed open another door that I could have sworn, I had not seen there before.

The second I entered into this cavern room, I felt my mouth just drop. It was beautiful. 'The Falls', as they called it, was a huge cavern area with a large pool being supplied its source of water from five medium sized waterfalls. What was extraordinary about it though, was there were actual rain droplets frozen in midair all around the room. I looked at Dawson, who gazed down at me. He smiled and offered me his hand. I took it easily.

He led me down the steps, just to the edge of the pool. I was surprised by how I walked through the raindrops, and did not get wet at all. It was as if they would shift or move away to let us through. Dawson sat down on a nearby step, while I took in my surroundings. I felt such a longing in my body, and such freedom. I raised my hands in the air and I started laughing. The droplets suddenly began to pour down on me, and it felt amazing. The water felt cool and refreshing on my skin. I started twirling around and opened my mouth, inviting the water inside. The taste of the rain in my mouth was stimulating. I was letting the rain wash away all of my pain, all of my confusion, and all of my weakness. I could not stop laughing and crying. The waterfalls began to pour more rapidly and there was thunder in the distance. Goosebumps began to form on my skin from the thrill of it all. I was having a storm, for myself.

I had forgotten that Dawson had been sitting there watching at me the whole time, not bothering to say anything. I stopped and stood there looking at him, a playful smile on my lips. He was smiling too, his black hair soaked and blue eyes twinkling.

"Why am I so happy Dawson? What is this place? I never want to leave…" I turned back towards the pool, dipping my foot into the lukewarm water.

Dawson rose and came up behind me, wrapping his strong arms around me. I turned around, and looked up into those beautiful eyes.

"You know, Trish I really shouldn't be doing this. I have to tell you that the thousands of years I have been alive, I have never experienced the kind of emotions that you make me feel. There is so much that you have yet to know about me, and even of yourself. It is forbidden for me to even want… you." Dawson said, moving a strand of wet hair out of my face.

"Why? Dawson, after all of this happened you were there for me and you found me. I honestly, do not care how insane or weak I may seem because you are letting me know that all of this is real. I want to thank you for that." I smiled up at him.

That made him close his eyes and he kissed my

forehead. His lips lingered there for a moment and the coolness of his breath on my skin, made me close my eyes too. I had figured he was never going to come back for me, so I simply tried to forget him. I would never do that again. The rain was slowly stopping, the last rain drops in froze midair like before. Dawson leaned away and stared at me, and then he moved his hands slowly up my arms, and removed the leather jacket I was wearing. I shuddered with excitement. I felt my pulse start beating rapidly. This was all new to me. My life had granted me a fair share of guys that I had ran off with when I had the chance, to escape my overprotective parents, but Dawson was surreal.

All of those childish, boyfriend memories that I had, came to an end when I met Dawson. This was real emotion coming from me, and I knew he had me. He was something that someone fantasized or dreamed about. Nothing about him was normal and he was nothing like other guys I knew. He acted exceedingly mature, yet he appeared to be handsomely young.

Perhaps it was his immortality, or his darkness that made him distinguishable from anyone else I had ever met. I hesitated, as he leaned forward.

"Dawson…" I began, but he placed his finger on my lips, and let his cloak fall to the ground, he gently guided me down the steps into the pool.

I was instantly aware that I was nude. How did that happen? The water was like acid, it had dissolved all of my remaining clothes and I had not even noticed. I looked at Dawson, who was shirtless, with only his black pants on. The water had eaten away our garments! I felt the butterflies in my stomach begin to flutter, and I started to feel self-conscious. My bruises! How could I have forgotten? I swam deeper into the water, trying to hide myself in the darkness of it, in the deeper part of the pool.

"Trish? Are you okay?" He asked, hesitating before reaching out for me. He swam out to me, and pulled me back to the shallow end.

I looked down ashamed, but to my surprise,

they were gone. There were no bruises on me. As the matter of fact, my skin looked flawless.

"I had bruises from Bryce, but now they are gone!" I said, astonished.

Dawson pulled me back into his arms. "'The Falls' has cleansed you from all impurities, inflictions, and flaws. You are as new as can be."

My adrenaline was soaring. It felt like being baptized. There was something in the water that was making it glisten like diamonds. All my anxiety suddenly melted away, and things in my lower body began to flare. I really wanted to do things that I probably shouldn't have. Our nakedness did not bother us at all and it surprised me.

"Trish, I want you to know that you are doing this. You are making the rain pour and stop, the water shine like crystals, and the waterfalls pour faster. I am merely reading your mind. You are such a powerful young woman, yet you have to learn how to control your powers for when you use them, Bryce can find you a lot quicker." He moved closer to me, pulling me into him. My body reacted to his touch.

"I want to know who I am Dawson. I do not want to be afraid any more. I have lost everything I have ever loved and I need to know what I have left to live for." I replied, caressing my face against his moist chest.

"You have more than you will ever know to live for Trish," he murmured, and with that he lifted my chin, and kissed me.

I felt all my shyness die away, as my hands glided over his muscular back. He brought me into him. I gasped as he began to kiss my neck. I wrapped my legs around his body tightly, and he pulled my hair back firmly. This made me kiss him harder, with more passion. He started kissing lower on my neck.

"Dawson, I need you right now. I do not know why, but I cannot control myself with you." I said, kissing his face all over.

"Trish, I am breaking a sacred rule with you,

which could cost me my throne, and you are saying it is you that cannot control yourself," he breathed into my ear.

We both laughed, then he drew back to look at me, and suddenly all of that loving emotion swept right off of his face.

"What is wrong? Dawson?" I asked, becoming worried.

"You're eyes Trish. They are silver. This confirms my honor and declaration. You are truly the Chosen One. We have to get you to my father immediately. When your eyes change color, it means that your power is building up to be put into use, and sooner or later you are going to have to unleash that power. It is part of what will become a daily routine for you, I guess you could say. Different emotions, events, or people could trigger your eyes to change color. I'm sure they already have done so before, but you haven't noticed. However, what I am worried about is that your power may release itself on its own time, because you do not know how to control it. This will have to wait, because if you cannot control your power then Bryce may feel it. We all can somehow feel it, and I will not risk losing you again." He kissed me hard and fierce one last time.

Confused, I just stood there helpless. What could I say, except that I felt like he blew me off because of my eyes changing colors? Okay. Maybe the situation was serious, but I still felt a little offended. He grabbed my hand, rushing me out of the water, before I could reply.

"Wait! What are you talking about? What about our clothes?" I yelled, as we ran.

Unhappy by the sudden closure and dumbfounded that we could not finish our moment together, I looked down and saw that I was dressed in a yellow gown with gold slippers. How was this happening? It had to be magic! We finally ran through an open door, into a nice dining area, looking as if we had spent the whole evening getting ready, and we were dry.

Looking around, I noticed a large oak table and Dawson's father seated there staring directly at me. The room was rather large and there were guards lining each of the three exits out of the room. All of them had bright, yellow eyes and guarded expressions. There were fresh fruits, assorted cheeses and meats, and what appeared to be three glasses of wine on the table. The tiles of the floor glimmered with blue and white sparkles. There were several portraits of most likely, family members, hanging along the stone walls of the room. The circular fireplace was lit and there sat a black purple-eyed cat, the size of a chair at least, on the rug. The cat was staring at me, with a glint of humor in its eyes. I did a double take of the cat. This was beginning to get to be too much.

"Father, we need to speak with you promptly. I am certain that Trish here is the Chosen One...." Dawson was cut off by Mavus.

"I know exactly who she is son. I felt her power

just minutes ago. Do you think Hejeena has made this an easy journey for me child? No, she has not. Now please be seated so I can discuss these matters with you. By the way, you look more presentable." He pointed to the chair next to him.

I nodded and sat down. Dawson attempted to follow, but his father raised a hand at him.

"Son, I would prefer to speak with Trish alone. I will inform you when you will be needed." Mavus said, strictly.

"Yes, father I understand," Dawson replied, taking his leave and never taking his eyes off of mine.

Chapter Five

I was not aware of how nervous I was until I realized my wine glass was empty, and I had not touched the delicious food that was sitting in front of me. Mavus had not said a word to me for the past five minutes. He just stared at me with those yellow eyes, as if trying to read me. He issued for the server to pour me another glass of wine and I did not protest. Finally, he cleared his voice and I shifted in my seat, uneasily.

"Trish, welcome, I rather do not know how to make such a tremendously long story short for you, but I find it relevant that you understand that nearly everything you have been told by Bryce or Dawson about your life, is unfortunately true. Hejeena and I carry an agreement that has been a burden with many of my relationships with the other rulers, and now that the cause of this agreement is sitting right in front of me, I am not pleased." Mavus spoke, harshly.

"I did not ask for this you know! I was perfectly fine with my ordinary life. And I did not ask to be a part of some mythological world either okay, so do not take your anger out on me! I do not deserve this!" I yelled at him, shocked by my own defensiveness.

"Compose yourself child!" Mavus sat back in his chair wide-eyed. I must have taken things to a higher level. I knew I was being disrespectful. I put my head in my hands and closed my eyes, miserable. The tears that fell down my cheeks felt like icy rain. I lifted my head and saw blue droplets smeared on my hands. Was I crying blue tears now? This could not be happening.

"I think this would go more lightly if you actually spoke with your mother yourself." Mavus replied.

"Yeah right, I can only imagine she is any worse than you are." I muttered, taking another drink of wine.

"Oh she is my dear, she really is, and you will realize that you have so much in common with her dear. I almost feel as if I am seeing a youthful reflection of her." He smiled, weakly.

"Is she your sister?" I wanted to make sure I got this family tree figured out. Otherwise, Dawson and I were in some deep trouble.

"We are hardly related child. All of us rulers were created by our great Father creator, and were coined as siblings, but we are not from the same bloodline. Each of us has created our own bloodline, and we are many. We are bonded as rulers with specified powers and duties and created by the same creator, our Father, which is why we treat each other as some sort of family, shall I say. We each had to create our own bloodline or our own family that would continue to rule when the present ruler's time was up. Along with our own family, we each created our own civilization. The time is almost here, where all of the kings pass their thrones down to their sons and the queens follow with their daughters. The most powerful son or daughter is next in line for the throne. That is how we work in these worlds out here, Trish. We are not related, but bonded. Sometimes we are the best of allies, and then at other times we are the worst of enemies. Nowadays, it seems like I have more enemies then allies, thanks to your mother." Mavus sighed, and closed his eyes tiredly.

I ignored the "mother" comment, and finished my second glass of wine.

"Your question wouldn't have anything to do with your sudden connection with my son Dawson would it? If you don't mind me asking," Mavus suddenly asked, opening one eye at me.

"No, not at all," I replied, coolly.

A little lie never hurt anybody, right?

"Good, because I want you to know that Dawson is the next in line to rule the Moon and I would not want him to rule with anyone other than the lady he is arranged marry. You are off limits to him and I would certainly be angry if he was to fall in love with Hejeena's most powerful daughter. Now that we have an understanding, I will let you be led to your room for the evening and tomorrow we will go present yourself to your mother. I have nothing left to discuss with you," he said, rising to stand.

I stood too.

"I would prefer to go back to Chicago and be with Gabriella, she needs me, and I do not feel like I belong here." I protested, the knot in my chest growing tighter.

Mavus's disapproval of me was unnerving. He did not even know me, yet he was so judgmental. How could he be against me and Dawson? Dawson was the only guy I had truly felt anything for and let alone could help me get through all of this sanely. He had protected and saved me. If I had to live in this world without Dawson, I would not be able to make it.

As if reading my mind, Mavus spoke, "You are to stay here for the evening and see your mother tomorrow. Then from there, you are on your own to decide, with your mother, what will be a proper arrangement for you. I will have another son escort you and assist you for the night. I have a moon to tend to and I will be keeping a watchful eye over Dawson. Good evening, Trish." He said my name with complete dissatisfaction.

With all that being said, he left the room, and the cat followed, glancing back at me smiling. I didn't believe cats could smile, but this one surely could. I wanted to go mad, but all I could do was stand there. How dare he do this to me! What was so forbidden and wrong about me and Dawson? Why did they feel that I was so powerful, when I only felt weak? The door that Mavus exited opened and in walked a tall slender man with long blonde hair that was tied in a braid. His eyes were the exact same color as Mavus's and he was handsome. Everything about him was golden, even his skin. He was wearing a black leather jacket and faded, blue jeans, carrying a helmet. He was a biker, I thought.

"Hey, I'm Griffin, one of the sons around here.

If you would follow me, I can show you to your room." He offered a dazzling set of white teeth with his smile.

I hesitated. I really had no other options and I might as well have jumped in a cold shower and realized that this was real. He turned as if he already knew I wasn't going to put up a fight and began to walk back out the door. Reluctantly, I followed him into the dark hallway.

Griffin stayed a few feet ahead of me, pausing now and then, so I wouldn't lose him. Throughout all of this, I had realized that both Dawson and Griffin appeared to be different from normal human beings. They had unique physical features, walked faster than a normal person would, spoke elegantly, and seemed to be stronger than average…a lot stronger even though they did not look the physique in appearance. Most of all were the eyes of these people, they all changed colors.

The corridors I walked through were filled with

complete blackness with a wall candle here and there to provide a faint dim light. I had to half jog to keep up with Griffin's pace. I was thankful for the fact that he was blonde because the candlelight reflected off of his hair making it easier to see in the dark.

"Here we are Trish," Griffin said, coming to a stop at a large wooden door.

"How do you know my name?" I asked, following him into the dark room.

He laughed. "Everyone around here knows your name by now, Trish. Many have awaited your presence for quite a long time now."

I did not say anything back. I was too annoyed by all of this commotion about my return. As far as I was concerned, I was just a simple Kansan girl with an ordinary life until my mother had to go and date a monster. When the room was lit my mouth dropped wide open. The room was the most lavish room I had ever seen. There was a huge canopy bed in the middle of the room draped with gold silk fabrics with lush pillows and golden covers. A large vanity was placed at the far side of the room with pure gold brushes and packed with powders and perfumes. The floors were black tile and there were two entrances to a closet and a bathroom. Moon and star lamps hung from all over the ceiling. I felt like I had entered some kind of enchanted room.

"It is a beauty is it not?" Griffin said.

"Yes, I love this room."

"Well that is good because it has been yours for centuries."

I looked at him astonished.

"Are you serious? I cannot believe it! How could a place like this belong to me?" I asked, in bewilderment.

"Hejeena has kept it nice and furnished over the centuries. She knew that one day you would return and she wanted you to be safe… with us," Griffin spoke, moving around the room and touching things here and there.

That did not make any sense because I was only nineteen. She must of known I would be born and had made prior arrangements, but had to protect me and send me away. Now that made sense.

"I will be assisting you from now on, well that is, until you meet your mother."

I sat down on the edge of the bed and looked up at Griffin harshly.

"Why am I forbidden to be with Dawson right now?" I asked.

I was starting to get angry and that was not good, especially not now.

"I cannot speak about this issue Trish," Griffin replied quickly, moving towards the door to leave.

I grabbed his arm.

"I demand to know! I have a right to know, Griffin. This is not fair at all! Please, I need answers!" I pleaded, gazing into his yellow eyes, so similar to those of his father's.

He tensed under my touch and he pulled away

immediately, as if he were in pain. He grabbed his arm and then took off his jacket.

"What's wrong?!" I asked, concerned.

"You burned me." He winced, taking off his leather jacket and placing his hand over a reddened area on his other arm.

"I did that?! How in the world did I burn you? I have never done anything like this before to anyone. I am sorry!" I went towards him, but he motioned me away.

"Look Trish, you are powerful and you need to accept that and work on controlling yourself better. I am keeping my control, but you are pushing the limits. I cannot bodyguard you if your power keeps unleashing at the wrong times," Griffin explained, calmly.

I nodded my head, trying to control my emotions. That would help right?

"I'm going to take off for a bit, but I will be back later to check on you. There's a cell phone that you humans use, for you on your vanity, my number is three on your speed dial. Use it if you need me, but otherwise you're in for the night. Hejeena left tons of gowns and accessories for you in your closet. If I were you, I'd dress impressively to see your mother tomorrow for the first time."

I rolled my eyes as he closed the door behind him. I heard the lock latch shut. Great, I was locked in my own room. Was I that dangerous or did everything have to remain such a secret? I looked around the beautiful room and tried to find something that could help me understand who I was. I came across a chest that was sitting in the corner of the room. The chest was large and covered with diamonds. I plopped down in front of the chest, forgetting I was in a gown, and winced as I heard the side of the gown rip wide open.

"Damn! This is just not going to work out." I said, standing up and heading towards the closet.

I was hoping this mother of mine would have some normal clothes. Perhaps, a t-shirt and some jeans would be nice. I doubted it. I opened the door to the closet, and it was as if I entered another room entirely. There were gowns hanging up all the way to the back of the closet where a large mirror stood. On the other side were arranged shoes, or should I say heels and slippers, and other accessories hanging up.

Any woman would have dreamed for a closet like this, me, well I personally would have been flattered, but not tonight. Out of the hundreds of dresses, I finally chose a simple purple gown. The ends of the gown flowed out freely and had a less likely chance of getting ripped or ruined. I put on some purple flat slippers and tied my hair back. Looking in the mirror, I felt like I was dressed to go to a prom. Memories flooded back into my mind. The life I used to know, the normal life, didn't exist anymore.

I went back to the chest and opened it. Inside were a book and several files of papers. There was also an embroidered red bag in there. I gathered the files, the bag, and the book and sat on my bed. I began to look through them. I opened the first file and was surprised to see my named typed across the first page. I skimmed through the file and began to realize that the pages were filled with information regarding my childhood. It even had photos of me. The other files were the same, each file held information and pictures of me all the way up to the past year. Was my mother spying on me my whole life and I never knew it? Shaking my head, I closed all the files, not wanting to relive the past memories of my life. I decided to look at the book instead. It was a large book with thick gold pages and a blank brown cover. There was not a title on the front so I figured it must be something someone would write in like a journal… a big journal.

I was almost correct. The book was handwritten and on the first page was a message. It read:

To The Chosen One…

Long before you were born your destiny had already been made. Your presence has long been awaited but not so your fate. Within these pages lies your past and what may become your future. The war you must fight and the path that you must choose is only for you to determine. The darkness may be your enemy but it is your only weapon. Realizing what is real and what is not is vital to your survival. You carry a power that is stronger than any ruler. Immortality is a gift, which is not promised, and can be taken away. There was not a soul who knew who you would be born from, but all of the planet's rulers knew you were coming and were eager for you to be theirs. I have kept my identity secret from all of my kin and it is you I have created for a special purpose. They refer to me as Father of Eternity and I have written and provided this for you Trish. What you have in your hands is your guide, your power, your life. May your journey be rewarding, not destructing. Farewell, my child.

I closed the book with goose bumps running along my skin. I opened the book again and the writing was no longer there, instead, there was a picture of a man. This man was really pale and his shirtless chest was broad and muscled. His jaw was wide and masculine and his eyes were pure black. He had long black hair, longer than Dawson's, and he was remarkably beautiful, but dangerous looking as well. He had and evil smile that displayed sharp fangs peeking out from full lips. I looked towards that bottom of the page and read aloud to myself in disbelief.

"Bryce is a Fifth Born, the son of Xandran, and the next in line to rule the Underworld. He is the prime danger and worry of all of the rulers for it is he who will lead the war for universal destruction and confiscation of all of the planets. Be wary, Trish." I closed my mouth shut and I felt a chill in the air.

How could that be Bryce?! He was an entirely different person as far as I was concerned. I turned the page and my answer was given to me. There was another picture of Bryce, the lawyer, the fake, disguised Bryce, the one who had killed my family. It made perfect sense for him to change his appearance, because if anyone saw him for who he really was on Earth there would be problems.

I closed the book again not wanting to read any further, but I knew this book held the only answers I would get about anything and everything. For some reason it seemed that every time I opened the book it was full of nothing but blank pages. I opened and closed it several times only to receive blank pages. Did this book only show me something when it wanted to? I opened it again and to my relief there was writing.

This book responds to your mind with the hidden answers to all of your questions. Think of what you would like to know, open the book and before you it will show.

I smiled and shut the book. Overall, there were so many things I wanted to know, but I was trying to organize them in my head. Then I remembered a part in Bryce's description saying he was a fifth born. What is a fifth born? I closed my eyes, took a deep breath, and then opened the book.

The Fifth Born

Each ruler of the planets, the moon, and the sun, gives birth to children, several children. Yet, when the fifth child is born, it is that child that is the most powerful of them all. The Fifth Born of each ruling family will be the next in line to reign over their planet. Normally after the fifth child is born, the ruler decides to stop having children; some have more though others do not. Normally the first four children are among the most powerful and the fourth is the second-in-command. It is often painful for the first four children to await the birth of the fifth child knowing that they are the ones next in line to the throne. Each child lives a regular human life and has all human characteristics and qualities until they hit puberty when their unusual features start to show. Their powers are also developing a lot faster as they grow older and at the age of twenty-one they are given immortality and their powers are reaching their peak. A good question, commonly asked is if everyone is immortal eventually then why are some of the rulers elderly and others not?

As I mentioned earlier immortality is a gift and is not to be taken advantage of. There are several things that may damage ones immortality. These involve how some use their power, the time when the next ruler must take over, leaving the option of the former ruler to live the rest of their life as human on earth without any remembrance of this life, or join me, The Father of Eternity, in my realm forever. Immortality lasts until it is time for next ruler to reign and that is why some rulers await centuries until they have their fifth born, for when that child is ready to take his or her place in the throne it is time for the past ruler to make the decision I mentioned earlier.

There are many stages to immortality and there can be punishments. For example, Mavus is elderly. He has been the ruler of the Moon since its creation and he is preparing to come join me. He is one of the greatest rulers I have created and his immortality has taken its toll on him from all of his power usage. He is worn and ready to pass on his throne down to his son. Mavus is a wise ruler and has lived for centuries. His aging simply came along within time. However, Hejeena, your mother, looks rather young possibly thirty. She has been around as long as Mavus, but she has kept exclusive limits on her power usage and splurges on vanity. Immortality all depends on how the user decides to manage it.

You are the fifth child of Hejeena's and also the Chosen One leaving you with more power than any present existing ruler.

I closed the book, incredulously, and lay back against the pillows. Exhaustion was beginning to take over me and I still couldn't fathom all of the new information that was registering into my mind. This was happening way too quickly. There was just one more thing I wanted to know before I called it a night.

"Why are Dawson and I not allowed being together?"

Dawson is Mavus's most powerful son and fifth child. Hejeena and Mavus made a pact that Mavus's sons would serve as protectors of her daughters and nothing more. Mavus does not approve of his sons being in a relationship that he has not arranged and most certainly not one with a daughter of the one woman he truly despises, despite himself. Mavus sees you as a barrier to his son ruling the throne. He never wanted his son to fully love a woman to the point where it may distract him from fulfilling his duties and he does not approve of an equal for his sons.

Dawson's emotions for you have made Mavus feel threatened because none of his sons have acted this way before, especially not his most powerful sons. Mavus is one of the rulers that believed in a single leadership in his monarchy, he despises rulers, such as Xandran who would prefer to have a King and Queen ruling together; furthermore you have been chosen to determine the fate of all of the planets. I am sure Mavus is not the only ruler unhappy with the idea of a young woman that is not even immortal yet and inexperienced becoming the one they must depend on, protect, and follow for the safety and survival of the universe. Just be careful Trish, love can be dangerous and you are in danger deeply without all of your powers and knowledge being complete and mastered.

That was enough. I gathered all of the files and photographs except my special book and dumped them back into the chest. I took my hair down, letting my long curly hair fall down along my shoulders and back. Too sleepy to deal with any more surprises, I took off my slippers and crawled into the warm bed. As I lifted up the covers I felt the small embroidered bag that I had forgotten about fall to the floor. I bent over, picked it up and decided to see what was inside.

My fingers felt something cool and stone-like at the bottom of the bag. I lifted a long gold, chained necklace out of the bag. Hanging at the bottom, in the center, was a huge blue stone that sparkled with a faint light. I was instantly filled with calmness and peace.

As I looked closer at the large stone it was as if I was looking out at all of the galaxies and planets. A shooting star went by and I realized that it was like looking through a telescope. I could see all of the planets in orbit with each other. It was beautiful. I fastened the necklace around my neck, the stone feeling slightly heavy and cool on my skin. If the calmness I was feeling was coming from the necklace then I would wear it every day. I needed to be calm. A small rolled up piece of parchment paper was in the bag also. I unrolled it and in gold cursive writing it said that the necklace was a gift from my mother and Mavus. That was kind of them to get me a gift together, especially when they seemed to hate each other.

It seemed that I had slept for only a minute when I heard my phone buzzing on the nightstand.

"Hello?" I yawned into the phone.

"Trish, it is Griffin. Just finished doing my rounds, and decided to call and check up on you. Sorry if I woke you." He said, not as if he truly meant it.

"Now that I think about it what time is it, and how long has it been since I've been here?"

"It's been about a week now and it is the early morning hours. That is why my shift is done," he responded, with a huge sigh of relief.

He must not like his job that much. I sat there with my eyes blinking through darkness. A week! It only felt like a few days. Gabriella would certainly have filed a missing persons report for me by now. She was probably worried sick.

"Trish? Are you there?"

"Uh, yea sorry, but I really need to go back and see Gabriella. She is like the only family I was starting to have again. She is probably really worried right now." I replied.

"Right, well I will speak to my father and make negotiations. For now you are due to meet your mother. Go back to sleep and I will go fetch her number for you to call her. I will be back at your room in the morning. Bye." He said, and hung up.

"Bye." I replied back to the dead line.

I did not think I was going to get used to this and I certainly did not know how to save a whole universe, especially when I did not even know how save myself.

Chapter Six

It seemed again only an hour after my eyes closed from the phone conversation with Griffin, when I heard soft footsteps inside my room. My eyes shot open as a strong hand came across my mouth. My scream muffled.

"Trish it's me, Dawson. I need you to not say a word because I could get into deep trouble being in here, but I needed to see you," he whispered in my ear, making sensations in my body grow and my pulse speed up.

I nodded acknowledging his words and his hand slowly left my mouth and cupped my chin. I could not see him in the darkness, but his scent was comforting and I felt the moistness of his lips as he brought them down to mine. My fingers wrapped around his body, pulling him closer to me, so that we were inseparable. He attempted to resist and pull back, but I would not let him.

"Trish, let me go," he said, his breathing

uneven.

"Sorry." I released my hands from him.

"You look beautiful, yet so much more like your mother now," he said, frowning and running his fingers through my hair.

I rolled my eyes.

"Dawson, we need to talk. I was going through my files and some other things… I am beginning to understand some of this, but I am forbidden from being with you. I'm sorry, but I cannot bear that. After I meet my mother and tend to Gabriella, I will speak with your father about this." I said, firmly.

He sighed, laying down on his side next to me and put his arm over me.

"Trish, I do not know why I cannot seem to control my feelings for you. Unfortunately, there is a lot that you do not understand nor do you know about the situations out here, but do know this, despite what my father believes or wants, I have fallen in love with you." I could make out the gentleness in his blue eyes as they glowed in the dark.

I smiled and cuddled into his chest and he kissed my forehead. He lifted the necklace on my neck.

“I see you found your birthstone. No wonder you are acting so calm,” he said, rubbing his fingers across the stone.

I smiled and snuggled closer to him. I was actually in a peaceful state of mind.

“Do not leave me Dawson, ever.” I said, dozing off.

“Trish, we have eternity and if that is how long it takes to fight this battle so be it. Sleep. There is a long day ahead of you.”

I smiled and fell into a peaceful slumber.

I woke up with emptiness in my stomach. I opened my hand and saw that Dawson had placed the crescent medallion in my hands once again. He must have retrieved it from the clothes I wore when I first arrived. I put it on the dresser and looked around the closed room and did not see any sunshine. It certainly still felt like evening. Dawson had quietly left while I slept. I yawned.

My cell phone buzzed.

"Hello?" I answered, sitting up. I was terribly hungry.

"It is Griffin and it is exactly six in the morning, that is in human time and your mother expects you promptly at nine. She is very anxious and has arranged a last minute masquerade to be held for your arrival. I am sending a servant with your breakfast and outfit she has chosen for you. Understand rulers of nearly all the planets will be attending this event. It will be a very important moment for you and everyone else." He said.

"Wait, a masquerade? You mean that I will be wearing a mask?! Has she lost her mind? I would prefer just meeting my mother alone. I am not ready to see all of this." I objected, heatedly.

"I do not think she cares what you or anyone else desires, Trish. However, I do have Gabriella's number. I have sent it to you via text and would suggest you save it to your phone. You are subjected to one important law… Gabriella is not to know anything about your whereabouts or what is taking place now. Simply let her know you are okay, but nothing else. Goodbye." And with that he hung up.

The line was dead and I was still holding it to my ear. Gabriella! I hung up the phone and opened the text, saving the number at once. Just before I punched the numbers into the phone, the door opened and the black cat with the purple eyes walked in with a wooden table cart following behind him. There was a deliciously cooked meal placed on the cart consisting of juice, muffins, eggs, and fried potatoes. There was a large glittered bag on the cart too with a red ribbon tied around it.

"Trish, here is your breakfast and attire for the event. Make haste." The cat purred.

"You can talk?! I do not freaking believe it!" I exclaimed.

The cat laughed.

"You are quite far from what I expected the Chosen One to be. We have waited such a long time for this," he mumbled, shaking his furry head.

"Okay kitty back off," I said, irritated.

"Oh my mistake, I truly did not mean to offend you. Perhaps I could introduce myself. I am Lunar, Mavus's head assistant," he said. There was a sudden blast of purple smoke and before me stood a man, laughing.

I look at him dumbfounded. This magical nonsense had begun to spook me out. He was tall and pale with short black hair and his purple eyes looked lilac. He smiled and his teeth still resembled those of a cat's. He looked to be about thirty, but I figured he was probably centuries old with the rest of them. The black tuxedo he was wearing fit his, smoothly. He had claws for fingers and everything about him seemed awkward and mysterious.

"Well nice to meet you Lunar." I replied,

kindly.

"How do you do madam?" He bowed.

I laughed.

"So, are you going to be present at this party that this lady is throwing for me?"

"Remember now she is your mother and of course my dear, but by all means do not call it a party. Your mother would be quite unpleased if you called it that because this is a huge event. You, the Chosen One, have returned to take your place among the stars," he said excitedly, pushing the wooden cart up to me.

"My place among the stars, you have got to be kidding. Everyone has lost their minds if they think I am the right person for this." I complained.

He smiled and turned for the door.

"Actually, you are losing your mind to think that you are not the one for this, see you there!" Lunar giggled and closed the door behind him.

Snapping back from my interruption, I picked up my phone again and called Gabriella.

“Hello?” Gabriella answered, sounding tired, but hopeful on the second ring.

“It’s me…” I said, closing my eyes, waiting for her screams to come.

They came.

“Trish! Where the hell have you been? I’ve been losing my mind over here, smoking packs of damn cigarettes every day! I went to the police station and they said you never existed and that I need to take my crazy tantrums somewhere else before I get into trouble. Can you believe that?! No records of you! A Trish Noble doesn’t exist! I even brought them your I.D.’s from your purse to prove it and they said it was all fraud! Who the hell are you Trish? Why didn’t you tell me you were going to get kidnapped and taken away? I am so lost and angry…Even though I really do not know you it was as if you were becoming my best friend.” Gabriella said, and I could hear the hurt in her voice.

I felt terrible. What wouldn’t I give to be there

right now to hold her and tell her the truth? Instead, I had to come up with something to tell her. Wait, did she say? I didn't exist?

"Gabriella, there's so much to it, but I cannot say…"

"Don't you dare hide anything from me Trish, I've done worried my ass sick over you and I'll be damned if this kidnapper is holding you hostage. You tell me right now where you are and I'll come save you and kill the damn fool with or without the damn cops," she said, furiously, interrupting me.

A small smile came across my face. Gabriella would do it too, I knew she would.

"Look, I'm okay, that's what you need to know. I will come back to you soon. I do not know why the cops have no records of me ever existing, but I am Trish Rosalyn Noble. I am real and I think my records have been stolen or erased. I am unsure of everything at this moment and I cannot talk long, but I do not want you to worry about me anymore. You mean a lot to me and I want you to know that I am somehow involved in some deep serious issues. Do not panic Gabriella, because I'm fine." I urged.

"Trish, please talk to me… Don't leave me hanging." Gabriella sighed, distressed.

"I got to go… I'll talk to you soon ok?" I said.

"Trish..."

I hung up. That was way too hard and I knew if I stayed on the line any longer I would probably tell her the truth. The smell of breakfast finally took its toll on me and I hopped out of bed and helped myself. I felt like I could eat the world. Piling eggs, hash browns, and a muffin onto my plate with a glass of orange juice, I sat down and began to eat. I was halfway done with my food, when I remembered the glittered bag my "attire" was in. I put my plate back on the cart and grabbed the bag.

It was quite heavy for just a dress. I pulled out the dress and gasped. It was long and gold with ruffles that went all the way out into a train. It was strapless and came with gold elbow-length gloves. I pulled out the gold matching slippers and my hand hit something hard. It was a medium-sized case. I opened it up to find a gold encrusted black mask with gold feathers. She was not joking, Hejeena was throwing a masquerade. I decided to go ahead and get dressed. I kept my necklace on to keep me calm, just in case my anxiety got out of hand. Everything fit perfectly, but I would keep the mask in its case until I left. There was a knock at the door.

"Come in," I said, anticipating who was outside my door.

A larger elderly woman walked in. She was wearing a brown gown and her gray hair was short and curled in at the ends of her face. Her large blue eyes were gentle. I instantly knew where Dawson's eye color came from. She carried herself with a grace that was full of gentleness and joy. I looked at her and the first thing that came to my mind was warmth. She looked at me with a motherly affection that I had not felt in too long. I wanted to hug her.

"Oh my, well hello dear, you look beautiful. I knew the dress would be perfect," she squealed, grabbing my hand and twirling me around.

"My mother did not pick it out?" I asked, confused.

"Well, yes she most certainly did. I just gave her my best advice." She winked.

"Oh. Are you Mavus's wife?" I asked, out of curiosity.

"Funny of you to ask, but yes I am. My name is Noreen," she answered, adjusting my dress.

I shifted, uncomfortably, as she tightened the dress straps from behind. Talk about not being able to breathe!

"I need to do your hair and makeup now because we have an hour before your departure." She hurried me to the vanity.

Noreen was such a talkative person. As she did my hair and makeup, she rambled away about the masquerade, my arrival, and everyone that would be there. She was such a nice woman and I wondered how she and Mavus came about to be married, because Mavus was certainly not a nice man.

"There you go! You look absolutely stunning my dear. I hit the bull's-eye every time," she said, victoriously, spinning me around to look in the mirror.

"Wow…" I managed to say.

My hair was in an up-do of curls and the gold eye-shadow brought out my hazel eyes. The bronze blush and ruby lip color made me appear enchanting. I actually thought I looked beautiful. Leaning closer to the mirror, I smiled.

"You look just like your mother." Noreen said, confidently, smiling.

My smile vanished. I looked just like my mother, great.

Griffin suddenly appeared out of nowhere and stopped abruptly at the door.

"My goodness, you look beautiful Trish."

“Thank you.” I said, noticing he was dressed in a black tuxedo and his long yellow hair was in a braid that went all the way down his back.

He turned to his mother. “Mother, if you and Trish will both come now, your entourage awaits to escort you to Venus.”

I badly wished he was Dawson, but I reluctantly took his offered arm. Noreen followed us. I took a deep breath and prayed. I had forgotten my book! As if reading my mind, Noreen hurried back and grabbed my book off of the bed and gave it to me.

“I figured you would like to keep your diary. I used to love mine when I was little,” she said, winking.

I nodded, knowing the book was far from being a diary and continued walking arm in arm with Griffin down the hall, and into the room where Mavus and I had our discussion the previous evening. Mavus was standing by the table in a silver robe holding a staff. His eyes brightened at me as I walked up to him with Griffin.

“Trish, you are absolutely beautiful. Words cannot even describe. Noreen well done my lady,” he said, and Noreen went to stand beside him.

“Do not forget your mask,” Griffin said, placing it on top of my book.

I took my arm from his, curious by how he had brought it, and held both the book and the case to my chest.

“So are we leaving now?” I asked, anxiously.

“In a moment, my other sons and daughters should be here in a minute. We are leaving by the portal, because we do not have much time left,” Mavus said, rather nervously.

Right after Mavus spoke a set of doors opened and in walked at least nine people. Three were females and the rest were males. I spotted Dawson and my heart started beating faster. He was beautiful. They were all wearing black tuxedos, but Dawson somehow stood out. His hair was tied back and his blue eyes met and held my gaze. He smiled at me with admiration. I looked at Mavus's other children, they were all too remarkably beautiful. Two of the girls looked to be the oldest and were all blonde-haired with blue eyes similar to Dawson's. Three of the boys looked to be in their early teens. Then the youngest, a little girl, was a little image of her mother. There was one other, besides Dawson and Griffin, who appeared to be around my age. He was much taller than Dawson with black hair and gray eyes… He looked really sad for some reason. Those gray eyes looked at me for a moment thoughtfully and then looked back down at the ground. I looked back at Dawson, who was looking from me to his brother, back and forth, frowning.

"My children, shall we?" Mavus said, raising his staff in the air.

There was a flash of light and the next thing I saw was a huge circular opening. I could not see through it, it was just like a large mirror, all we saw were our own reflections. Noreen went first followed by the other children, except Dawson and Griffin, into the portal. It looked like the portal just absorbed them into it. I was astonished to watch the mirror suck them in so easily. I wasn't sure I'd be able to do it. Dawson stood there, waiting.

"Aren't you coming?" I asked him.

He looked away from me, clearly upset.

"Dawson is going to be arriving with his fiancé. He is awaiting her presence now. Come on Trish," Mavus said, with a triumphant smile on his face.

My face fell. Had he just said fiancé? That bastard! Dawson had a fiancé? Oh boy, did he have some explaining to do!

"Trish, listen…" Dawson began.

I refused listen. Acting on impulse, I grabbed Griffin's arm and walked right into the portal without a second backward glance. When I walked through the portal three things happened at once. I felt an electric shock, I lost grip of Griffin, and I was pretty sure I fainted. Wherever I was, I definitely knew it was not Earth. I felt like my skin was on fire.

"Trish! Griffin will you pick her up! She has fainted. I knew it probably was not the best idea to bring her through the portal for the first time. The sandstorms always cause problems. The electrical current is off, causing the portal to sidetrack. Hurry, through here," Mavus yelled, over the hot wind.

I tried to blink my eyes, but they instantly filled with debris from the air. Griffin held me tightly to his chest. I was amazed by how I appeared weightless to him, especially with the dress I had on. I managed to get a glimpse of another portal and we entered it quickly. Now I could breathe. Griffin set me down on my feet. I held on to him to catch my balance. Where were we?

"We are about to enter the masquerade Trish. Your mother is waiting for you. Here is your book and mask. Put it on now, we are about to enter." Mavus whispered.

This was all happening too fast. I put on the mask and took in my surroundings. We were in a small elegant room with stone benches and statues. In front of us was a huge castle door that descended hundreds of feet into the air. This was remarkable. I knew I should have been nervous about meeting my biological mother for the first time, but I wasn't. I was just tired and ready to get it over with.

All of Mavus's children, Noreen, and Mavus himself were wearing masks now. I saw Mavus grab Noreen's hand and I thought of Dawson. How could he hurt me? He should had been there holding my hand and walking in with me to greet my mother. Shaking my head, I tried to regain my focus on my mother. I looked down at my book, tightening my hold on it. The doors before us opened slowly.

Soft melodious music was being played by an orchestra. There was a red carpet that led to the throne. First I saw hundreds of people or what appeared to be people, standing and gawking at me. All different colors these beings were, some even looked alien. They all had masks on, but I could see the opened mouths and hear the whispers and bickering above the music. There was too many of them. Even the necklace could not keep my nervousness away. The room was so enormous. There were rose petals and candles everywhere. Servants lined the doors and walked the floor with platters of food and drinks. This was surreal, I thought to myself.

"Trish, begin walking toward your mother, pay no mind to the others around, we must walk behind you, now go." Noreen whispered in my ear, giving me a light push.

I glanced behind me and saw that Mavus and Noreen were positioned behind me and his children were behind them. They stared at me, waiting. Embarrassed, I turned and focused my attention ahead of me, on my mother. There was a tall woman sitting in a royal chair ahead of me. The second our eyes met, I knew this was the woman whom everyone said to be my mother. Everything about her was me, well at least genetically. We both had the same long curly, dark hair, hazel eyes, beige skin, and tall hourglass frame. I swear we could have passed for sisters or even twins. I held my breath as I walked closer, for I was truly frightened by her presence. She did, however, look much older than me with her crow's feet and fine lines on her face. Something about her was scary though. There was a powerful vibe about her. An inherent darkness engulfed her and her set facial expression made her appear to be a force not to be reckoned with.

Hejeena stood with a forced smile as I approached the stairs to the throne. She was wearing a long red gown that hugged her perfect figure and she wore glistening ruby jewelry. I just stood there. Her red lips pursed together in sudden irritation. She must have wanted me to go up there and hug her. There was no way I would be doing that anytime soon. I looked from her to the other four girls, sitting like queens, in the chairs beside hers. They all looked like flawless dolls. If they were my sisters, it would make sense to why they were glaring at me with disapproval. I was far from perfect. Yet they knew that I was the one. I resembled Hejeena more than any of them did. Noreen took my book from me and gently shoved me forward, reassuringly.

"My daughter, Trish, welcome home!" Hejeena boomed, walking down the steps and pulling me into her arms.

She held me too tightly. It began to hurt. Why wasn't I hugging her back? She smelled strongly of roses and I felt I could drown in her scent. I placed my arms around her slowly, and I heard a roar of applause from behind me. She let go, but remained holding my hand.

"Future Kings and Queens come and pay your respects! Here is your new Queen of Venus and the one we call the Chosen One before you all. Let the ball begin!" Hejeena exclaimed excitedly, pulling me to sit in the chair she had been seated in.

The second I sat down, my sisters' rose and walked away appalled for whatever reason. None of them looked like Hejeena, two had red hair and hazel eyes and the other two had light brown hair and brown eyes. I did not understand why they were all wearing red dresses with red masks, but I didn't ask. If Hejeena cared at all by their sudden departure, she did not show it. She sat down next to me, as people began to dance and enjoy themselves. A lined formed in front of me of all the future kings and queens of the planets. They were eager to meet me.

"Oh splendid, darling, this is Blaze and his wife Sonya from the Sun. They always bring the best gifts. He still has a few millennia to go as king. He is quite handsome, it always annoyed me how he ended up with such a plain woman like Sonya," Hejeena said in my ear, as the first couple approached.

I fought the urge to look at her and yell at her for everything she had done to me and put me through, but I decided not to. This was not the place and time for all of that yet. Why did she expect me to be comfortable with all of this? The sound of her voice annoyed me and I found myself trying to lean away from her intoxicating breath. True to her word, Blaze was really handsome though. He looked Spanish with his black hair, brown eyes, and darker complexion. He was dressed in a white buttoned up shirt with black slacks and knee high boots that made him look like a real bullfighter from Spain. In comparison his wife was not as beautiful as him, but she was pretty in a natural way. She had long red hair tied back in a braid with freckles sprinkled across her face and orange-brown eyes to complement them. She was dressed in a white gown that showed off her petite frame. I smiled as they approached because they were adorable together, in my opinion. They took off their masks and bended down on their knees. That made me uncomfortable.

"Trish, it is a pleasure to meet you finally," Blaze said, planting a kiss on my hand.

My hand felt like it was slightly burned, but it felt amazing, like the warm feeling of the Sun on my skin.

"Nice to meet you Blaze and Sonya. Sonya you look beautiful." I said, sincerely.

"Why thank you Trish!" Sonya said, and I could tell that compliment meant a lot to her.

"Please accept our gift from the Sun. I know you will take great care of him," Blaze said, snapping his fingers.

I expected it to be a puppy or something when he referred to "him" but to my surprise it was a blue baby dragon that was revealed to me from under a blanket by a servant. My eyes widened. Was this real? The little dragon stirred in his sleep, a bit of smoke blew from his nostrils.

"Oh my goodness, he is darling!" I said, hopping up from my seat, and picking the rough, scaly dragon up. I felt like a little kid at a toy store. Holding the dragon was like holding an extremely large lizard. Hejeena looked alarmed by my immediate interest in the dragon and how I just crouched down in my dress without a care to hold it.

"That was not quite the gift I was expecting for her. Trish will you get back in your seat please," she said, coolly. I ignored her.

"Thank you so much! I will take care of him. You might have to give me a few pointers on it, but I'm excited. I've dreamed of having one but never in reality believed it could happen or that they even existed." I said, sitting back down with the sleeping dragon in my arms.

"Well we are glad you like him. Every ruler ought to have their own personal dragon. He will protect you. Once again it is our pleasure." He nodded at me and my mother and they walked away.

My mother had a servant take the dragon from me and my book from Noreen to my room. She accused me of complimenting Sonya to make her mad. I shook my head. What was her problem? It was my first time here and she was already acting as if I was some sort of big disaster. This was not going as she had hoped, I could tell. Before I could ask, a man with purple skin, yes purple skin, came to stand before me. How much more weirdness would I be able to handle?

"I am Pradon, the new King of Pluto. Greetings Trish." He spoke slowly, bowing his bald head. His voice was deep and raspy. I stared at him with wide eyes. His eyes were completely white and that was it. No irises or pupils, just plain whiteness peered at me. The black cloak that hung on his frail body looked heavy but he wore it effortlessly. No one was with him and I wondered what his kingdom looked like.

"Nice to meet you Pradon," I replied.

He smiled a sharp toothy grin and took his leave from the throne. I looked at the long line in front of me and figured that it was going to take forever to get through all of the kings and queens, but it went faster than I had thought and I received several gifts. I was looking at an orb that Claire the Queen of Neptune gave me, when I heard a soft grunt from my mother. I looked up. Dawson. Great, this was just what I needed.

He approached me cautiously, holding a young beautiful woman's hand. I ignored the urge to push her away from him. Dawson left her a few feet behind him and came and stood in front of me. I looked up at him, waiting.

"Trish… We have already met. This is Margaret," he said, motioning for the pale blonde-haired woman to come closer.

She shook my hand firmly and smiled, but it did not reach her eyes. Then again, neither did mine. He must have mentioned something about me to her. That or we just sensed each other's competition. I could not seem to understand why she had possession of the man I loved, but with a bigger reality at hand, I could not allow myself to dwell on it.

"We would be grateful for you to accept our gift from the Moon," he said, reaching inside his pocket.

"I think I have quite enough gifts. You may keep yours." I said, firmly.

He glared at me and then nodded, pulling his hand out of his pocket. My mother looked at me in shock. Margaret looked suspicious.

"Trish you are never to reject gifts from other rulers," she said, clearly disappointed in me.

"Then you keep it for yourself, as a goodbye gift. I am parched, if you'll excuse me, I need a drink." I said, sarcastically, rising from my chair.

Hejeena stood quickly and grabbed my arm.

"Trish you cannot just leave, you have not met everyone yet and you still have to choose the leading male for your dance," she insisted.

"I do not feel like meeting anymore people right now. I am tired and want to take a break."

"I would prefer you have your dance first, it is an important part of the ball."

I was going to keep the argument up when Dawson's brother, the one I had noted earlier with the sad grey eyes, came up to us.

"Sorry to interrupt, but Trish do you mind if I

be the man for your Reigning Dance?" he asked, giving me a glass of what appeared to be wine.

Even though it was not what I wanted to drink, I gratefully drank it to the last drop. My mother looked at me with embarrassment. Now she thought her daughter was an alcoholic, what else was new? I looked at him closely. There was something about him that I could not say no to.

"Sure, I would like that," I said, my cheeks slightly red from the wine.

He smiled and I couldn't help but smile too. I forgot that Dawson was still standing on the throne with Margaret. Hejeena just looked relieved that I had accepted the dance offer.

"Well I guess that is fine, but you must wear your mask," Hejeena said, finally.

I nodded and began to return to my seat to get my mask when Dawson stopped me. He grabbed my hand and I yanked it away. He did it again, this time holding me roughly. I looked around at those observing us. This was horrible.

"Trish, do not dance with Luke, I mean it," he said, through gritted teeth, trying not to look upset.

"How you dare tell me what to do. I will make my own choices." I was surprised by the sudden authority in my voice. I've never acted like that before and I was beginning to wonder if all of this commotion was getting to my head. Hejeena glanced curiously at both me and Dawson. Dawson stood back hopeless, and Margaret came up and grabbed his arm. She clearly wanted to leave.

I snatched my mask furiously, put it on, and returned to take Luke's arm. Luke smiled at Dawson, wickedly and then moved his arm around my waist. Dawson looked away. I was a little uncomfortable with that, but let it go. If there was a jealously feud between the two of them, it wasn't my problem. My mother went to the throne and announced to everyone that the "Reigning Dance" was beginning. I looked out at the crowd and couples began to take position on the floor.

Then they turned their attention to me. They were waiting on me. I was confused. Luke patted my hand.

"Are you ready Trish?" Luke asked.

I looked back at Dawson, who hadn't moved since I walked away from him. Margaret was glaring at me and trying to get Dawson to go with her to the floor. He kept his eyes locked on me and Luke. Nervously, I looked up at Luke.

"I do not know how to dance," I said, trying to back out.

He leaned forward and whispered, "Don't worry, the magic does the dancing," and he kissed my cheek. I did not exactly know what that meant, but I smiled and nodded anyway. He pulled down his black mask over his eyes and his gray eyes twinkled at me. We began to descend down the stairs and onto the floor.

Chapter Seven

Luke held my hand tightly as we moved to the center of the floor. We got positioned like everyone else and then we waited. The lights dimmed and the soft melodious music began to get louder. Yet, no one danced. Luke pulled me closer to him.

"Why isn't anyone dancing?" I asked.

Luke smiled at me.

"Look down," he said.

I did and gasped. Sparks were coming up around our feet from the floor. I looked at the other couples and the same thing was happening to each of them. Suddenly, almost involuntarily, our bodies began to move with the beat of the music. We started ballroom dancing and I was not even sure I was doing anything. I looked at Luke and I knew what he meant by the 'magic doing the dancing'. There were tiny little pricks of electricity running along my skin. It felt exhilarating and I started to laugh as we danced, as we all danced. All the couples seemed to be enchanted and lost in the moment as well. I looked around in a daze. I did not want the dance to end.

“Trish, your eyes are silver,” Luke said, just moments later.

I froze, losing any trace of the excitement I had just seconds ago. Dawson had gotten afraid when they were silver before. Was this a bad thing? What was I supposed to do? Why were they silver? Something inside of me changed. I felt strong and powerful. It scared me.

“Do not fight it Trish, it’s your power embrace it,” Luke said, holding me tighter.

“I don’t know how!” I yelled, feeling suffocated by how tight he held me.

Pushing Luke off of me became impossible. I started to get really hot and my adrenaline started to rise. Sparks started to turn into flames around me and Luke. Surprisingly, the flames did not hurt us. I looked around us and everyone had stopped dancing to stare, everything was quiet. I began to see everyone for who they truly were. These creatures and beings took off their masks and stared at me with shock and fear. There were too many different faces with alien features and horrifying expressions gawking at me. Where was I? What were these beings? Nausea came up in my throat. I saw Mavus rushing towards me with his staff in the air and then I saw my mother doing the same. What was going on? Dawson, I needed him! I looked around and couldn't find him anywhere. The burning sensation in my stomach started to grow. Venus's kiss was having no mercy on me. I tried to push Luke away, but it was like he was clinging onto me for dear life. We started to rise into the air.

"Trish, this feels amazing. I cannot feel anything. I'm numb," Luke said, as if he were drugged.

His gray eyes were turning even grayer.

"What is going on? What am I doing?!" I demanded. Beads of perspiration formed along my forehead.

"Your power Trish, you are embracing it. Just

calm down and control your anger or you will hurt me. You were happy dancing with me and your power was unleashed by your happiness. You are feeding off of my energy and deep inside you love it. Do not be afraid of it or it will change, because as soon as you fully master it you will be so full of power that you could have anything you could ever want," he said, smiling, as if he were dreaming about it for himself.

He was right, it felt good, but as I looked down at all the others, they looked scared. For some odd reason, the look of fear on their faces excited me. Hejeena and Mavus were standing there before us with their mouths open. They were holding hands. Why? Were they trying to stop me? As if I was going to let them do that. I looked at Luke and wanted to touch him, to kiss him, to have every breath he took. I began to lean in towards him. I could tell he was afraid now too. He resisted, but then surrendered. Just before my lips met his, I heard a yell over the crowd.

"Do not kiss him Trish! You will kill him!"

Dawson yelled.

For a brief second I glanced at him and then I smiled, maliciously. He was running through the crowd right at me. His blue eyes turned to yellow. Was he going to attack me? I did not care, because I suddenly knew I could take Dawson down if I wanted to. I no longer felt like I was myself. All I wanted to do was suck the life out of Luke. I turned back to Luke, an alluring smile upon my lips. As if I were possessed, I ignorantly kissed him.

"No!" I heard too many yell at once.

I opened my eyes, instantly, as I was kissing Luke and felt as if my lips were glued to his. I was suddenly drowning in power. Not a soul existed to me, except Luke's. He began to pull away, but he was too weak, and I was enjoying it. It was as if I were taking a journey through his life. I kissed him deeper, more thoroughly. The pleasure of his pain was like biting into a forbidden fruit, irresistible. Luke's eyes closed and he fell to his knees. I went down with him, not taking my lips from his. His color began to drain from his face as we settled on the floor again. I knew he was dying and I wanted his soul. The next thing I knew, something threw me off of him, hard. I hit something really solid, which was probably a wall. I felt the impact of the wall for only and second and then white and black dots scattered across my vision. I was knocked out cold.

When I opened my eyes, my vision was blurred, but I could make out my mother, Mavus, and Dawson peering down at me. I felt like me again. Thank goodness!

"What happened? Where am I?" I said, attempting to sit up, but failing. They motioned for me to lie back down.

"That was quite remarkable Trish, for a beginner, I have not seen anything like that in my whole existence. You almost killed my son Luke." Mavus said, giving me a peculiar stare.

He appeared to be still in shock. Killed Luke?

Me? I was dizzy.

"Trish, we have had to made everyone leave unfortunately, Dawson insisted on staying here with you, against our wishes of course," Hejeena said, and quickly glanced at Mavus, then continued, "You are in your room, but need time to rest, Luke will be fine. He is healing right now. I have to go now and explain myself to all of the other rulers why the Chosen One is unable to control her powers and try to reassure them that we are not in immediate danger. Oh, how fate can curse some! Just rest and we will talk about this later."

She kissed my forehead, forcibly, then headed toward the door, grabbing Mavus's arm. I knew she was afraid of me now and did not want to accept me as a daughter and have another incident like this.

"Dawson," Mavus warned.

"Leave us father," Dawson said, angrily.

The tone of his voice made my eyes go wide. I rubbed my temple with my hands. Reluctantly, Mavus left the room with Hejeena.

When we were alone, I positioned myself against the pillows on the headboard. The room was too dark for me to really see it clearly, or maybe I was just dizzy. I did make out a stone wall and a fireplace with a chair next to it, which Dawson was sitting in. His tuxedo jacket was off and his white collared shirt was slightly unbuttoned. His hair was down and fell across his shoulders, perfectly. I wanted to go to him and run my hands through his hair, but I didn't, or more like couldn't. I felt weak and tired. Dawson was silent, staring at the fire with his hands clasped together against his chin. What was I supposed to say?

"Dawson…" I began.

He cut me off.

"Trish you almost killed my brother."

"I did not mean to! I do not know what happened. I am sorry… I," I was interrupted again.

"Why did you dance with him? What is your problem, why can you not just listen to me and what I tell you to do?" He turned to face me in the chair, angrily.

Was he serious? Now I was getting upset.

"Why didn't you tell me you had a fiancé huh? How do you think I felt!?"

"Look Trish, you think I planned that? It was my father's decision and the arrangement took place centuries before I even met you. Margaret is a daughter of one of the nearby moons. Her mother and my father planned for this engagement to happen, not me. I know you know that I care about you, but you have to understand that this is what is best for my kingdom. If it makes you feel any better, I do not love Margaret. I protect her because she is the future queen of the Moon." He drew in a deep breath, displeased with me.

"This is not right Dawson. I wanted to dance with Luke and I did not do it to make you mad. It was like something about him drew me to him. The power I felt was beyond my wildest imagination. I do not know how I came about to do that, but I knew I wasn't myself. I would never try to kill anyone just for the pure enjoyment of it." I explained.

"Well, I apologize that I had to hurt you, but I had to save Luke. You can become very dangerous Trish. Everyone was frightened because they've never seen someone younger than twenty-one express their powers in the way that you did. We have to understand that you are the Chosen One. This type of behavior jeopardizes everyone. We have to get you trained. Now that you have done this, Bryce is surely planning on capturing you within the next few days at least. Your power extended far beyond the galaxy with just that action alone. I cannot lose you Trish." Dawson rose and began to walk towards me.

"But you cannot be with me either." I groaned.

He stopped walking.

"Not quite. Please do not make this difficult for me Trish." He pleaded.

Upset, I forced myself up, ignoring the pain in nearly every muscle, and began to look for something to put on that was not a dress. I rummaged through drawers and the closet, but found nothing. Dawson just watched me. I sat on the bed and put my face in my hands, hopeless. I felt strong arms hug me from behind. I turned and put my face into his chest, miserable.

"I brought these for you. You left them at that hotel in Chicago," he said, bending down by the bed.

I looked at what he picked up and it was my bag of clothes. I was surprised, but did not question how he retrieved them. Never in my life had I ever been so grateful to see a pair of jeans and a t-shirt. I quickly changed into them not noticing Dawson had picked up my book.

"What is this?" He asked, opening it.

I quickly jumped on him, making him fall back on the bed. I grabbed the book from him and kissed him lightly on the lips. He chuckled.

"Nothing," I replied, smiling, mischievously.

I placed the book back on the nightstand and he looked at me suspiciously. Then he pulled me back onto the bed with him. He brushed a strand of hair away from my face and placed it behind my ear. For a good few minutes we sat there silently gazing into each other's eyes.

Unable to help myself, I brought up the subject of Margaret.

"So, why did your father choose Margaret of all women?"

Dawson moaned, like he knew this was coming, then rolled over on his back and stared at the ceiling. I almost thought he wouldn't say anything, but he did.

"Well, all of the moons of the planets put on a huge moon festival every year to honor the Father of Eternity. My father is the ruler of the most powerful and commonly known moon. That evening I was introduced to so many women descended from these moons. None of them captured my interest and I therefore, turned them all down. My father kept pushing me to find a bride because time was running out for when I was to become king. I told him I would prefer to rule alone, but he was not having me break our family tradition. Finally, he introduced me to Margaret. She was very beautiful, charming, and the most powerful daughter of her moon, Io. She was my father's ideal pick for a woman, my mom liked her, and I just gave in. We've been together for at least one hundred years, but there is never a day I wake up and feel that she is the one for me and we both know that. It hurts me to see her care so deeply for me and I cannot feel the same. She wants to hurry and proceed with a family and I cannot give these things to her." Dawson explained.

"One hundred years! That is a long time Dawson. My longest relationship was nearly five months. You are with someone you do not even love because you are afraid of disappointing your father. You are living a lie." I said, sitting up.

"You do not understand Trish. I must abide by my father's rules. He was the ruler before me and knows what is best for the kingdom and if an arranged marriage keeps the kingdom together, then I will have to do it."

"No you do not and if you must it should be with someone you love. Mavus clearly does not believe in equality. I know I have not learned enough about this, but I know that even if you were to rule your kingdom dominantly, you will be an unhappy ruler forever." I said, looking at him.

He sat up, looking into my eyes.

"I love you Trish, but I am not in my right mind right now. I cannot afford to mess up and neither can you," he said, lightly kissing my lips.

Deciding that I was not going to get through to him, I changed the topic.

"Why was I drawn to Luke?"

Dawson glared at me. Obviously it was not the best subject to bring up either. I shrugged.

"Luke is second in command, Trish. He is the rebel in the family and we do not know what to do about him. Luke and I have never gotten along. He is always gone and never fulfilling his duties. He travels to Earth often, putting himself in the dangerous position of exposure, and spends the evenings in bars. Despite everything, Luke has the power to charm women and he does it quite often. I want you to stay away from him. He normally targets vulnerable women and sometimes he accidently kills them, because he falls in love too deeply. His immortality bothers him for some reason. He does not understand that he cannot kill you, for you are not like mortal women, but you will kill him, which is probably why he is drawn to you." Dawson said, turning to look at the fire before he continued, "Regardless, he is still my brother, and I must protect him. I know he would not do the same for me, but I love and honor my family."

I did not know what to say, I just held his hand.

Dawson obviously didn't want to talk anymore about the subject of Luke or Margaret. Instead, he positioned himself on the bed leaning on his elbow.

"You need to get some rest and eat. The end of January will be here before you know it and from what I've been told you have a long few weeks of training ahead of you. I feel bad because I threw you off of Luke pretty hard, but since your powers are getting stronger, you should be healed by tomorrow." Dawson said, pulling me closer to him.

"I have to do training? Do you mean like martial arts or something?" I asked, playing with a strand of his hair.

He laughed, "No, your sisters will be training you in relation to your powers and how to use them. By the end of the month you should be able to control your powers a lot better, but since you are the Chosen One, it will be a lot harder to train you, because after all you have more power than the majority of all rulers, but you do not have all of your power yet. That makes you all the more vulnerable to Bryce's plan."

"And what plan is that?" I asked.

Dawson looked away from me as if he were

envisioning something.

"Everyone knows this by now, but Bryce plans to make you his queen. Once he marries you, by oath, he will have access to your powers. That's just how their world works and he will most likely possess you, so that you two will destroy the earth and take over the planets and galaxies together."

I could not believe that this was possible, but with the way Dawson looked at me, I knew he was worried about this. A knock at the door interrupted our discussion. Dawson was immediately up and at the door in a blink of an eye. He opened the door and Margaret stood there, with her hands on her hips, glaring at him and me angrily. I sat up and glared right back, because there was no way I was going to let her walk all over me because she was engaged to Dawson. Suddenly, a vase from my dresser flew towards me, but Dawson grabbed it before it could hit me. My eyes widened.

"Margaret enough," he ordered.

Had she just done that? Was she trying to hurt me now? I was about to speak, but she got to it first.

"What are you doing in here with her Dawson? Everyone left this pointless masquerade and I have been waiting for you with your father for quite a while now. I am upset that you are here and I demand to know why?" She pierced him with her angry blue eyes.

"Trish had been hurt and I had been attending to her. I also had a word to her about Luke," he glanced at me, "and I was just leaving, so if you will calm down, we can go."

Margaret became calm right at that moment and gave me a fake smile.

"Well, Trish I hope you do get to feeling better and maybe next time you'll be wiser not to mess with Dawson's family," and with that she grabbed Dawson's arm and yanked him out of the room.

Dawson shot me a "sorry" look and closed the door.

"What the heck! I already had enough to deal

with and now I have to deal with his jealous fiancé. Gosh, life sucks!" I yelled out loud to myself.

I went into my bathroom and washed all of the makeup off of my face and put my hair into a bun on top of my head, some of the curls poking out, but I didn't really care. When I went back into my room Lunar was sitting on my bed. He was in his cat form and it looked like he attempting to make a disgusted look with his face.

"Whatever are you wearing child?" He purred.

"It's what we humans call a t-shirt and jeans. My favorite outfit," I sneered. I was not in a good mood, but I hopped on the bed next to him. "How did you get in here anyways?" I asked.

"I have the ability to transport myself places," he explained.

"Well what do you want? Did Mavus send you here?" I demanded to know.

He smiled. "No I came on my own. After all, I had to make sure the Chosen One was well after the incident that took place earlier today." He began to clean himself. Gross.

"Will you people just stop with the Chosen One crap and just call me Trish? I hate sounding like some mythological being or something."

"Sure, well, Trish I should be going now, I will see you soon," he said, his purple eyes looking at me.

"Wait, you are Mavus's assistant. There is something you aren't telling me, why are you really here?" I was getting annoyed again.

"Oh no reason, I already told you I was just checking up on you." Lunar replied, stretching out his paws.

"Seriously, do not act like you're a real cat, okay."

He glared at me, smirked, and then vanished. How rude! I wanted to ask my book some questions, but I was too tired and I snuggled under the covers and thought about Dawson, my life now, my life then, Gabriella, Lunar, Luke, and Bryce. When I didn't have the energy to think anymore I was sound asleep.

Chapter Eight

I awoke to someone calling my name and nudging me. Pulling the covers off from over my head, I looked up to see one of my sisters standing there. She was one of the redheaded ones with hazel eyes like me. Her hair was in a bob cut, her makeup done perfectly, and she was wearing what a red skintight bodysuit.

"Wake up sleepy head! We are waiting for you at breakfast. I am going to be training you today. By the way I'm Sherry, second in command." She was bubbly and energetic. Sherry stepped back from my bed and I noticed how the bodysuit outlined her thin, but toned silhouette. How could someone look that perfect? It was not motivating. I moaned and pulled the covers back over my head. There was no way I was getting up. I was too tired. Suddenly, the covers were thrown off of me and I was yanked out of bed.

"Let me go! What do you think you are doing?" I said, irritated with her and freaked out by her strength. I mean she had just pulled me effortlessly out of my bed.

"Goodness, what are you wearing?!" Sherry's large hazel eyes looked me over from head to toe.

"Oh, leave me alone already." I yawned and wiped my eyes.

"Mother's going to be quite unpleased with you," she said, shaking her head amusingly.

"Oh well, what else is new around here." I replied, following her out the door grumpily.

The interior of Hejeena's kingdom, as everyone was calling their residences, I would just say castle, was definitely something that looked like it got its look from Greek architecture. Outside of my room were hallways border with marble railing that formed a rectangle around the second floor that we were on. As we walked around, passing several closed doors, I was able to see down to the first floor, which was where the masquerade was held the previous day.

A marble staircase led us down to the first floor and the throne was smack down in the middle. None of the tables, the red carpet, or the servers was there from the previous night. There were just guards lining the entrance and decorative statues everywhere. It was very Greek traditional looking with the stone columns and architraves surrounding the space. One of the statues was of Aphrodite, the Greek goddess of beauty and love. I followed Sherry past the throne and through a huge set of double doors.

Hejeena was sitting in a rather uncomfortable looking stone chair and there were platters of various breakfast foods placed on a long stone table. My other sisters were sitting there as well. They all stood when I entered and then each of my sisters let out giggles and snickers. Hejeena stared at me as if she'd been slapped. She quickly recovered, stood, and placed her hands on her hips. My sisters silenced and sat back down in all of their beauty and grace.

"Trish. It is foolish of you to come to breakfast in that form of attire! Where did you get such articles of clothing? I made sure not to place anything of that kind in your closet." Her ruby lips formed a hard line, and her long curly hair fell down her back in voluptuous spirals.

Everyone was dressed elegantly. Even Sherry looked nice in her training outfit.

I sighed, knowing that I could never live up to their expectations. "Look this is what I like to wear. I am comfortable in it. Never in my life have I felt pressured to be something I'm not and I'm not going to start wearing ugly gowns and slippers just to please you." I said, putting my foot down.

My sisters widened their eyes, surprised by my reply, and looked at their mother waiting for her response.

My mother glared at me. "Oh no, you will not speak to me that way ever Trish and from now on, until you are twenty-one and immortal, you will do what I say. You will not make me regret having you as queen."

"Where I'm from we become adults at the age

of eighteen and according to the Father of Eternity I am the Chosen One, whether you like it or not." I said, not moving an inch.

She looked as if she would rip my eyes out from where she stood. With pursed lips she said, "No one speaks of the Father of Eternity around here young lady and I do not care what you were brought up to believe, you little earthling! You are my child, my daughter, and you will listen to me and adapt to our ways. You are no longer on Earth and never will be. No more of this discussion, after we eat and your other sisters introduce themselves you will begin your training." She poured herself a glass of wine and sat down, her breathing even again.

Reluctantly, I went to the table and sat down. Sherry sat down next to me and the other girls looked as if they wondered why I hadn't been punished for my outburst. Hejeena's hand shook, with anger, as she lifted the glass to her lips.

"Well do not just sit there speechless girls!

Introduce yourself to your sister!"

One by one my sisters began to introduce themselves.

"I'm Rachel, it's a pleasure." The other redheaded one said.

She was wearing a dark green gown that brought out her hazel eyes and her red hair was long and curly in comparison with Sherry's. Rachel was what one would call cute, not pretty. She even had the freckles to add to the cuteness. She sat back down without looking at me anymore.

"My name is Madison," said my sister sitting next to her, her tone deeper than usual.

Her long straight, light-brown hair was tied in a French braid. She held my gaze with large chocolate brown eyes with long eyelashes that anyone would envy. Madison appeared to have some sort of cream colored draped dress on that wrapped around her whole body leaving her shoulders bare. She was the only one not wearing makeup, but no one would be able to tell with her flawless complexion. We smiled at each other and then she went back to eating.

Finally, I looked next to her, where my last sister sat. She glared at me and rolled her eyes before talking, putting down her drink. Her personality reeked of Hejeena's influence.

"Tracy," was all she said.

It was obvious she resented me more than the others and was also the most beautiful one. Her wavy light brown hair had a little blonde to it and flowed down past her shoulders. Tracy's brown eyes were smaller than Madison's, but more alluring. She had high cheekbones, a supermodel body, and had great posture. It almost looked like she was carved from stone. Her lips were full and red, but they weren't smiling. She was dressed in a black strapless gown and had long black painted nails to match.

"Staring is rude you know." Tracy said, cutting a piece of an apple.

"I was just observing that's all." I replied,

raising my eyebrows, surprised by her coldness.

"Well, maybe you should observe yourself!" She snapped.

"How about you, stop being such a bitch!" I shot at her.

She stared at me with her mouth opened.

"I bet you have heard that language before haven't you?" I said, reaching for a bagel.

"That's enough you two!" Hejeena commanded, clearly fed up with the discussion period.

The others paid attention to their food, but I saw Sherry smile slightly from the corner of my eye. Victory was mine I thought and smiled as I bit into my bagel.

"Now, Trish, you are the youngest, then Sherry, Rachel, Madison, followed by Tracy. I expect you all to get to know each other and develop close relationships, because eventually you will all need each other." Hejeena, said, picking at her salad.

Who ate salad for breakfast? I did not know

exactly when I would ever accept them as my family. Even with Hejeena, I had a difficult time calling her my mother. Despite the odds, I was stuck with them. I was miserable and they knew it. We either would become bonded or we would drive each other insane. All women, no boys, and I suddenly thought of little Henry and my heart fell. I wanted my little brother back. Tracy began to speak and it brought me out of my thoughts.

"Oh please mother, I am quite disappointed. How did you give birth to her? Something like that?" Tracy looked at me disgusted.

I glared at her, stuffing a piece of the bagel in my mouth. One thing was for sure, me and her were not going to get along, at all. Hejeena shot her a warning glance, and then turned to me and smiled.

"Darling, Trish, if you will please watch what you eat. I expect all of you young ladies to maintain nice figures," she said, taking my other half of my bagel and putting an apple in its place.

"So you are going to starve me now! You know what?! I will die here, before Bryce ever gets a hold of me!" I stood, shoving my chair back and began walking out of the room.

"Good, that will make life for us much easier." Tracy muttered.

I turned.

"You know what Tracy, I could care a lot less about your life and if the day comes when you need me, if you keep up with your attitude, I just might let you die!" I said, coldly.

Her lips opened as if she would respond, but she didn't. Sherry rose from the table.

"I guess it's time to get to training, Trish," Sherry said, lightheartedly, trying to ease the tension. I followed her out of the room and knew that Tracy was shooting darts at me the whole way out.

Getting into my training suit had been a trip. Sherry had braided my hair and offered to apply my makeup, which I declined. I asked if she had any lip balm instead, but she had never heard of it before. Go figure. My suit was black and fitted to the point that I could not breathe. I felt like I was in a leather scuba diving outfit, minus the flippers. I was wearing a pair of black boots instead. To make it worse, I was claustrophobic. I followed Sherry out of my room and she led me through a portal that took us to a vast field. I could make out the sun and even the moon in the distance. It felt like I was still on Earth, but there was a thin layer of something, almost like a thin sheet of film in the sky, that extended across the field and made it look like we were trapped under a plastic veil. It seemed to me like we were in a glass ball.

"Hope you like our training spot. It is nice and private and goes on for miles. We can make illusions and practice magic unlimited her," she said. I followed her to a shed on one the side of the field.

"Remember, I told you I had a surprise for you." She smiled.

This brought my attention back into reality.

“What is it?” I was curious.

“She opened the shed and out came my blue dragon. How could I have forgotten about him? Too much had been going on. He had grown to be about my size over the past day! The night before I received it from Blaze and it did not make sense how fast he had grown. I smiled and laughed as he ran towards me, knocking me to the ground and licking my face.

“Wow! Didn’t you grow,” I said, getting up, and wiping the dirt off.

“Oh, I know! Dragons grow quickly. Give him another week or two and he’ll be like seven stories high.” Sherry said, patting him on his side.

“He will not speak to you, until after you name him,” she explained.

Speak? I went to my dragon and tried to think of a good name. It was hard when I wasn’t one of those people who thought about having kids or pets or spent my time daydreaming about baby names. The dragon was blue and had gray eyes. His scales shimmered with a mixture of silver and light blue. He was friendly looking, not evil, like many of the dragons I had seen in movies growing up.

“Moses?” I asked, looking at him.

He shook his head quickly. Guess he didn’t like that name. I tried to think and then I thought of it.

“I know, how about the name Draco?” I asked, smiling.

“I love it!” The dragon replied, in a low rich voice.

I was surprised by how human his voice sounded. Draco sounded just like a young boy. I laughed when he came over and licked my face.

“Draco, that is a constellation right?” An amused voice said from behind me. I turned to see Luke standing a few feet away with a tickled look on his face.

He had healed! As the matter of fact, he looked great. He was dressed in jeans and a fitted black shirt which outlined his perfectly ripped abdomen. His black hair was spiked up and his gray eyes were staring at me flirtatiously. He had his hands in his pockets and rocked back and forth on his heels.

"Yes, as the matter fact of it is. It means dragon and that is why I chose it." I replied, keenly.

He was attractive, too good looking for my taste and memories of kissing him the night before resurfaced.

"What are you doing here Luke? Right now is not the time. I need to train Trish alone." Sherry said, annoyed by his interference.

I forgot Sherry was even there, which proved just how distracted I was by Luke's presence.

"Sherry, you know everyone trains here. We just happened to drop in at the same time." Luke said, nonchalantly.

"Sherry, I did not know everyone trained here…" I looked at her.

"Yeah, I guess I forgot to mention that little detail. Except there are several training fields spread throughout the galaxy, not just this one, that everyone trains at." She didn't take her eyes off of Luke.

Draco growled at him. I did not understand the tension. If Luke was one of Mavus's sons, and they were the good guys, then why did Sherry and Draco not like him?

"Well, now you know Trish, so maybe now we can all get to our training. You don't mind if I join you do you?" He smiled a perfect smile.

"No of course not," I said. My head started to pound and I couldn't stop thinking about kissing Luke. Something wasn't right. I did not feel like my powers were acting out right now, but I could not be sure and I did not want to I risk it. I would not have responded to Luke the way I did either. He must have been messing with my mind. Maybe Luke had the ability to possess people's minds or at least play with them.

"Actually, since this is Trish's first training session, we'd prefer to train alone." Sherry intervened.

Luke scowled at Sherry. "Trish said it was

okay. Shall we train, now?" And before Sherry could respond, she was somehow thrown across the field and landed smack down on her back. I stood there in disbelief. She was back on her feet in an instant.

"Seriously Luke! You really want to do this," Sherry yelled at him.

She ran towards Luke with magnificent speed, her hands in fists and she punched him making him fall to the ground. He tripped her with his leg, making her fall, and then he kicked her hard in her stomach. Luke stood up, he smiled at me. I shook my head for them to stop. Sherry was still on the ground and out of nowhere Luke somehow shot a yellow fireball at Sherry, making her fly across the field and into the door of the shed. His eyes had turned yellow. I stared at them both with fear. Sherry was leaning against the shed with her eyes closed and blood trailed from her lips. Adrenaline raced through my veins. What was happening? What kind of training was this? I was suddenly furious at Luke for his conduct.

Luke turned to look at me, his yellow eyes piercing mine.

"Your turn," he said, coming towards me.

Draco stood in front of me, hot smoke coming from his nostrils. He had been quiet this whole time, but had never left my side.

"Leave her alone." Draco said, harshly.

Luke laughed. "Oh come on Draco boy, she has to learn how to fight some way. Do not worry I won't hurt her too bad. Besides, unless she commands you to, you cannot interrupt us. Draco, remember that you are not immortal and your life is not valuable. I will kill you."

That did it. I moved in front of Draco, my attention solely fixed on Luke. Sherry hadn't moved and was obviously hurt. Whatever training this was it wasn't right. There was no way I was going to let him touch my dragon or hurt Sherry again. This needed to end, now.

"Enough Luke, look I don't know what you are

trying to do or prove, but you need to leave now," I ordered, my teeth were grinding.

"Is that any way to talk to the man you almost killed the other night. You see I like you Trish. I like you a lot, but unfortunately, I prefer my revenge over lust, first. Maybe I can take you out on a date afterwards," he said, continuing to come towards me.

Draco growled, baring his huge, sharp teeth.

"Stay here Draco. Don't do anything." I demanded, and by the way he stood still, I knew he would obey.

I walked up to Luke so that there was about a foot of space between us. His eyes were still yellow. I did not flinch at all as I held his gaze. He smiled and swiftly grabbed my neck and threw me across the field with such force that I started to see black dots in my vision. Not again, I thought. I got up on my knees gasping for air. A fierce pain ripped across my stomach and Venus's kiss began to burn terribly. Draco whined as if he felt my pain.

"How did you like that for a first hit?" Luke said, walking towards me.

I stared at the ground feeling like I was going to faint. Then the strangest thing happened. My body began to get really hot and my adrenaline soared. All of the sudden I started laughing and I am talking about laughing to the point of tears and I felt no more pain at all. I just wanted to kill. It was as if something or someone had taken over my body again. I turned and looked at Luke, who froze to a stop a few feet away from me. He sensed my change and I knew he felt the power in the air.

"Trish? Your eyes, they are silver!" He took a few steps back, startled.

"What is it Luke, are you afraid? I won't bite too hard." I smiled, reaching out toward him.

"No, stay back, remember we are training Trish!" Luke exclaimed, backing up further.

"Are we? I thought you wanted to get revenge. You see Luke you cannot defeat me for I will kill you and take away your immortality. I will devour your soul and take every inch of you into the darkness. I know all of your secrets and I know all about those poor girls you killed." I couldn't believe I was saying all of this. Who was I? Where was I? I felt like I was balled up somewhere inside of my body and someone else was doing the talking. Luke looked scared.

"What are you doing to me Trish? Why can I not move?! Let me go!" Luke begged, looking like he was trying to move, but was stuck in place.

"Honestly I don't know what I'm doing, but I don't care. You are not the hotshot you thought you were huh?" I said, my voice sounded like I was singing every word.

I, truthfully, did not want to hurt Luke, but something inside of me did not care what I wanted. Sherry was starting to come back to her senses and Draco just stared at me, incredulously. I went and ran my hands up Luke's arms and he shuddered and closed his eyes at my touch. When he opened them they were that beautiful gray color again. I ran my finger along his chest and up to his lips tracing them. I loved how petrified he was and I brought my face up to him, tilting my chin as if I was going to kiss him.

"I love you," he whispered, involuntarily, and I let him bring his arms around me.

I laughed, it was a wicked laughed, and answered. "I know you do Luke. You love me enough to die for me don't you?"

"I do," he responded automatically, like a robot.

"Good, because I am going to kill you now," I spoke.

"Okay," he replied, softly, not protesting.

I saw Sherry coming towards me. "Trish do not do it, you are not in control of yourself right now. Try to control your powers." She tried to reason with me.

I looked at her and for a moment I felt myself get calmer and I thought I was beginning to control whatever was going on inside of me. Whatever had a hold on me was not having that though. Instead, my hand rose towards her and without my consent, a red fire ball shot out and hit her. She flew, hitting the ground with a thud. She was knocked out cold. I smiled and then turned back to Luke. He was now trembling.

"Did you not think I was capable of something like that? Now where were we?" I said leaning in towards him.

"Trish! What are you doing? What is going on here?! Stop now at this moment!"

I turned to see Dawson standing a couple yards away. His eyes were yellow and the muscles in his jaw clenched. He was wearing faded blue jeans. Dawson was barefoot and shirtless, his long black hair let down. I wanted him to help me, but I did not have control of myself. A part of me wanted to run and let him protect me and another part, well wanted to kill him too.

"Dawson, so nice of you to join us," I said,

grabbing Luke by the neck.

"Let him go, Trish now," Dawson ordered.

"Oh but darling, you were not here just moments ago. You should have seen what your dear brother Luke did to me. Since you did not see it, let me show you." And with that being said, I threw Luke across the field.

I turned back to Dawson and he shot a huge yellow ball of fire at me, hitting me right in my midsection. I fell to my knees holding my stomach. Dawson ran towards me, catching me before I fell. I reached up and touched my nose. Blood was dripping from it onto my fingers. I felt what was inside of me, or my power leave me and I felt as if all of my bones were broken. Dawson's eyes went back to their normal blue and held such pain in them. I touched his face, my blood smearing on his skin. He did not care. He laid me on the ground, gently.

"Trish, I am sorry. Forgive me. Every time I hurt you I feel like I could die. I hate myself." He looked away from me, like he was trying not to cry.

"You had to do that Dawson. Please help me. I do not know what is wrong with me. I do not want to be evil and I do not want to hurt anyone. I just want to be able to control myself and that was not me out there," I spoke, losing breath.

Draco came and lay next to me and I smiled at him. He made me think of Nikki and Henry and I knew that I would not let him down the way I had failed them. I could taste the metallic flavor of blood in my mouth. Draco looked away from me, upset. Everyone was upset. Sherry stumbled over holding her side. She looked terrible and her face was all bloodied and bruised.

"I think you should be the one training me," Sherry said, smiling, trying to hide her pain.

I could not help but smile. She was too optimistic, even in the worst situations. I wished I could be like that.

"Sherry, take Trish home and get her cleaned

up. Take care of her. Tell Hejeena that father and I will be over this evening for supper. We have to discuss some issues. Meanwhile, I am going to go deal with Luke." Dawson looked down at me and then at Luke lying across the field.

Without another word Dawson walked towards Luke. Sherry helped lift me up and opened a portal. Before we entered the portal I turned and looked back at Dawson and Luke, who was beginning to stand up. The last thing I saw was Dawson punch him sending him right back down to the ground.

Chapter Nine

When Sherry and I returned to the kingdom, Hejeena was waiting for us. Her face suddenly was filled with shock and concern.

"What in the world happened to you two?" She asked, worried.

"We ended up having a problem." Sherry explained, "Luke showed up and made Trish lose control of her powers again. Mavus and Dawson plan to be here shortly for our evening meal to discuss this matter with us."

I had never seen Hejeena look outrageously angry, but at that moment her eyes contained fire. She looked at me and then at Sherry and I swore fumes were coming from her nostrils. I just wanted to take a nice hot shower. My body was sore, but I knew Sherry had taken a worse beating than I had and she had not deserved it.

"So be it. I want you two to get cleaned up and ready for their arrival. Trish, I want you in a gown."

Hejeena glared at me, her expression daring me to protest.

I wanted to argue, but decided against it. I was too tired and I might as well get the night over with. The conversation was clearly going to be about me and my lack of self-control. Sherry walked me to my room and left to get ready. I looked at my book and I went to pick it up. I wanted some explanations before I confronted the others tonight.

"Why do I have a lack of self-control over my powers?" I asked myself.

My dear Trish, what a pleasure to hear from you! Quite disappointedly, you have not progressed as much as I had hoped you would for the first week upon your arrival. You have seen how dangerous your powers can be, however that is just a mere amount of the problem. Controlling your powers comes from within your mind. Let me put it this way, Bryce knows you cannot control your powers yet and there is a part of you that does contain evil Trish, for you are the

Chosen One. You are like a balanced weight of good and evil. The good side of you is constantly at war with the evil side of you. You have to decide which part will be stronger than the other. Each part benefits the other in a way. In this case though, presently your evil side is winning. Bryce wants that part of you as his queen. You must start with the basic mind control steps and that is learning to control your emotions and thoughts first. This needs to be done before Bryce gets a hold of you. He is not more powerful then you, that is not the problem. Bryce has mastered his powers and knows how to trick people into situations or deals they cannot get out of... Practice, practice, practice, my dear...

Someone knocked on my door, hard. Just as I closed my book Hejeena stormed in.

"Why are you not ready?" She snarled with her hands on her hips.

She was dressed in a white gown and her long hair was pinned up with diamond-studded pins.

"Well I was just reading," I began, but before I

could finish she snatched the book out of my hands.

"Give that back now!" I yelled.

"Reading? Reading, blank pages Trish! I have had enough of your nonsense. You are my daughter and it is about time you become more disciplined. Get ready and you will not be seeing this journal of yours anytime soon!" She slammed the door in my face.

I was angry, too angry. I felt like I was being treated like Cinderella. I looked into the mirror on the vanity table. Forget dinner, I was going to practice. If I could get my powers under control as my book said then I would be able to get out of this place and hopefully kill Bryce. I knew if I killed Bryce I could hand the kingdom over to Sherry and resume my life on earth with Gabriella. What would I not give to be normal again? Well, Draco, but I figured I would take him with me somehow.

Staring into my reflection, I tried to concentrate on my feelings. What could I think of that would make the mirror break? I thought of Bryce and Luke and more rage swept through me. The mirror started to rattle. I kept my focus on my anger and pictured the mirror as Bryce. The mirror broke. I let out a deep breath. Okay, now how could I put it back together? I thought of my old life, my family I loved, and the feeling of being in Dawson's arms. I stared at the broken pieces of glass hard and to my surprise they slowly lifted themselves up and went back into place. I lifted my hand, and laughed as a light flashed and sealed the cracks of broken glass back together.

I did it! My mirror looked as it had before. Someone opened my bedroom door.

"You are not ready." Hejeena stood at the door way, like a dark, cold ghost.

My smile faded. Why did she keep coming back every five minutes? It was irritating!

"I was just about to do that." I turned to face her.

She held a bright pink dress in her hand.

"What is that?" I asked, disgusted.

“This is what you are going to wear tonight.” Hejeena smiled.

“No I am not. I am going to pick out my own dress,” I said, walking towards my closet.

“Go ahead and try.” Hejeena said.

I gave her a look and opened the closet and shrieked. Everything was pink! How did she do that? I turned to glare at her. She smirked and tossed the dress on the bed.

“The dresses will stay that color for the rest of the night, so if I were you I would get dressed now. Would not want Dawson to see you looking that hideous would you?” She laughed, leaving the room.

“I cannot stand her!” I screamed, to myself.

“Me either!”

I jumped and turned to see Lunar sitting on my bed. His purple eyes smiling at me amused.

“Have a rough day?” He purred.

“You have no idea.” I replied, as tears formed in my eyes.

"I do not? You have no idea of the amount of bickering I had to listen to between those two brothers." He stretched.

"What are you talking about?" I demanded.

"Well, I really should not say, but Dawson warned that if Luke did not stay away from you that he would take away Luke's being second in command. Luke literally lost his mind. He told Dawson that he was just jealous that you prefer him and that you were the Chosen One. He told Dawson he would never be half the man that Mavus was for letting some "little skank" like you interfere. Worst of all, he stepped down from being second in command. Luke left for Earth." Lunar clawed at my blanket.

I sat there holding the pink dress in shock.

"Are you serious? He called me a little skank? Oh my goodness, this is terrible Lunar! What am I going to do? I do not even know what to say!" I said, with worry.

"Well go take a shower, I am sure you will

think of something in ten minutes when you are expected to be at the table or your mother will start a catfight." Lunar laughed.

"Oh, crap!" I frowned at him, he disappeared and I went and took a quick five minute shower.

I walked into the dining room wearing my pink fluffy dress, clearly upset. Hejeena looked at me and grinned. Mavus sat there in a gold robe, drinking wine, not really paying attention to me. Dawson was absolutely stunning, as always. He was wearing a black tuxedo, his hair down and his blue eyes staring at me. I felt ridiculous in the dress and knew Hejeena wanted to humiliate me, but I was not about to let her get away with it without taking her down with me. Before I sat down I bent down to tie my converse shoes that I decided to anger her with. It worked.

"What in the world are those?!" She said her tone high.

"Converse, they are my favorite. I took them with me to Chicago. Thanks to Dawson I have my favorite belongings here." I gave a big flashy smile and sat down.

She turned to Dawson, who looked down at his plate, a small smile on his face. I looked at Mavus, who looked at me disappointedly. My smile vanished and I coughed and straightened my posture. Mavus nodded, respectively and then Dawson did so too. We all cheered with glasses of wine. I quickly drank mine aware of them all staring at me. Then I started on my glass of water.

Mavus began, "Trish, I was informed of the incident that took place earlier today. For the second time you lost control of your powers and almost killed Luke, my son."

"Wait! I would not have if he had not…" I was cut off.

"Let me finish young lady. Luke should not be a problem anymore for he has exiled himself from his position and from the Moon itself. I am in agony right now Trish and for some reason I want to blame all of this on you, but I know that I would be wrong. Luke has made his own decision. Griffin is now second in command." Mavus stated.

"We have to take certain precautions now." Hejeena spoke.

"What do you mean precautions?" I looked at Dawson and he looked away.

"You are not permitted to leave these grounds under any circumstances until you have mastered your powers and until you turn twenty. Do not let this bother you, for time goes faster up here. Dawson found you in January. It is now mid-February. A few days here is almost a week sometimes, it depends, but bottom line is you do not turn twenty until late September. That will give us plenty of time to get you in top shape for your reign as queen." Hejeena said, slicing a piece of what looked like steak.

I groaned. "You are not serious! Look, I want all of you to know that I have made progress already. I broke my vanity mirror and put it back together just by practicing mind control!"

Hejeena choked. "You did what!"

"That is really great Trish." Dawson said happily, reaching for my hand, but Mavus stopped him.

"That is not enough Trish. You are still too dangerous. We cannot risk anything at this point. I have already lost one of my sons," he eyed Dawson, "and I am not giving you anymore of what belongs to me. I will be gone soon, but I am not leaving until this war is over." His yellow eyes held my gaze sternly.

"So what are you going to do? Lock me up in this torture chamber until my birthday? I am not going to do that and I will not." I said.

"You do not have choice." Hejeena snarled at me.

I wanted to smack her. This was not happening. Before I could argue the lights in the room started flickering on and off. Hejeena looked at Mavus and she became edgy. Suddenly the room was terribly cold.

Dawson was by my side in a second. He held me closely. Mavus and Hejeena stood and moved in front of us facing the entrance to the room. The guards that were in the room, about ten of them, went and formed a line in front of Hejeena and Mavus. There was huge gust of wind and the doors burst open. My worst nightmare was standing right in front of me in his real form.

Bryce walked into the room, all seven feet of him, broad and muscled. He was in a long, black robe, his long black hair braided and his black eyes staring at me as if I was the only thing that ever existed in the universe. He was sickly pale as if his body had not seen the sunshine for years. Then again it probably had not. As he came further into the room I found myself drawn to him. It was odd, but part of me desired to be in his presence. I found myself drowning in his black eyes and I could see that he wanted me. He wanted to own me. My body shook.

"You are not welcomed here Bryce." Mavus

said, coldly.

"Sorry to intrude on your little party Mavus, but I have endured long enough without my dear future wife." Bryce said, elegantly, walking closer towards me. His voice was like honey and I wanted to drink his words into my soul. What was wrong with me? Dawson noticed how I was reacting to Bryce and did not like it.

"Stay away from her! Stop whatever it is you are doing." Dawson held me tightly, his eyes bright gold yellow.

"Aw Dawson, I know you love my Trish, but have you forgotten you have a fiancé? You have your own queen, let me have mine." Bryce smiled, his sharp teeth glittering.

Hejeena went and stood in front of Bryce. She looked small and powerless compared to him.

"Trish is never going to be married to you! She is the queen of this planet and you might as well turn around and marry one of your walking corpses you breed down there!" She spat into his face.

Bryce laughed and my ears started to hurt. This whole time he had never taken his eyes off of me. And I could not break away from his spell. I knew I had to do something before this got out of hand. I did not want anyone to get hurt. I did not want to play hero, but then I realized maybe that was better than being the chicken. I would never hate on a heroic movie again.

"What do you want Bryce?" I said, surprised that my voice barely trembled.

"You already know what I want Trish. Leave with me now. I have waited long enough for you and you knew that I would come eventually." Bryce took another step towards me, pushing Hejeena out of the way.

"She is not going anywhere. Leave before this gets out of hand. This is not your territory here and you are outnumbered." Dawson said, through clenched teeth.

"Dawson, boy, now calm down. I did not come

here to fight or kill anyone. I simply came to retrieve what belongs to me. Trish, come now." He gazed at me lovingly with an offered hand.

"No!" Mavus, Hejeena, and Dawson yelled at him in unison.

He just smirked and sighed. Did he expect it to have been easy? No, he was playing around the bush, I knew it.

"Do you all not understand yet? Trish has not told you? I know she knows now. Come on Trish, why will you not explain to them about how you are bound by an evil form? We are basically two peas in a pod. It is our destiny to be together." Bryce smiled, reassuringly.

They all looked at me with confused expressions. That devil! I had to explain myself. This could be bad.

"I am not I swear. I am the balance of both good and evil. It is almost like there is a good part of me and a bad part, but I promise you that I can learn to control the bad part and the good will dominate me. Dawson, you have to believe me, you know the truth!" I grabbed him, "I am not evil and I did not mean to hurt anyone! Do not let him fool you!"

Bryce laughed, wickedly.

Dawson looked into my eyes. He was searching for answers.

"I believe you." He said kissed and me lightly.

Bryce hissed. Dawson's love for me was too strong for Bryce's persuasion tactics. He moved on to my mother.

"What madness do you call this Hejeena? You have given birth to a monster. I know you do not want her to be queen. I see your disappointment every time you look at her. Let her go where she belongs, among the dead." Bryce looked at her, persuasively.

Hejeena looked down and shifted uncomfortably, but then looked at me like she was considering what he was saying. I could not believe it!

Bryce turned to Mavus. "My friend, she almost

killed Luke, your son, second in command. She forced him to isolate himself in sorrow on Earth. He has lost everything because of her. Do you really think she can save the Universe?"

Mavus turned away from us ashamed.

He finally looked at me. "Trish, my dear you belong with me. Imagine the life we could have together. You could have anything you ever wanted and I do mean anything. I know you miss little Nikki and Henry terribly. I could give them back…"

"Shut up! I am not going anywhere with you. You are filling their heads with lies, my head with lies! I am not evil. It is you that is evil. I already know what your plans are and I am not going to let or help you destroy the universe and control the planets. You have taken everything from me! I am going to kill you myself. That is a promise." I spoke with such anger and adrenaline rose in my body as I forced Dawson to let me go and walked up to Bryce.

He watched me amused by my sudden

confidence in confronting him face to face. I knew I had control, because I felt I could unleash the power inside of me whenever I was ready. Control felt good. I could feel it slipping a bit, but I had a hold on myself. Bryce looked down at me. Bryce was frightening up close and I felt as if I had made the dumbest decision of my life by putting myself that close to him. I felt lightheaded.

"Trish you cannot kill me and you would not." Bryce lifted up my chin, effortlessly.

I yanked my face away. "I will kill you Bryce even if I have to die doing it."

"Well, we shall see about that when you turn twenty. You see you would not want to try anything unwise tonight Trish, my love, because that dear Gabriella of yours is in down there with me and if you do try anything she will die. Now I gave you the opportunity to take leave with me now and I will let her go," Bryce continued, "but if you do not come to me by your twentieth birthday, I am going to have to start what I plan to do. Destroy, kill, and conquer. My army is already ready, all we are missing is you, but if you do not cooperate I will be forced to go on without you, and kill you as well. My love, I do not want to do that."

He had Gabriella! What had I done! I had to save her. Mavus, Hejeena, and Dawson were staring at me and Bryce in disbelief and fury. I remembered Father of Eternity telling me of how Bryce used trickery and deals against people. He did not have Gabriella yet. I knew he was lying, that is what demons did. I decided to play along, let my evil side lurk to the surface just a bit.

"Wait, Bryce. Let's make a deal," I said, finally.

He smiled and leaned in towards me.

"Trish, do not do that! What are you doing! Get away from him!" Dawson screamed.

All of the sudden Bryce had my arm and we disappeared. When I opened my eyes we were in a dark room. I was sitting in a chair at a table with a candle lit on it and Bryce was sitting across from me. He looked even scarier than before with the light from the flame flickering shadows across his face. I sat there stiff and afraid, but I would not let him see that.

"Tell me about your deal Trish. Let us negotiate." Bryce's black eyes were filled with eagerness.

I was nervous, but the evil part of me was thrilled also. That evil part of me wanted to play games with Bryce and challenge him.

"Well, as I said, I will come to you on my twentieth birthday. However, you have to leave Gabriella alone and leave everyone else alone to reign over their planets as they wish. Their freedom for me to live, to rule, and to flourish forever," I said, watching him carefully.

"Done," Bryce answered, too quickly.

"Okay, take me back to my kingdom," I said, anxiously.

"First things first," he said, his dark eyes piercing mine.

"What?" I said, pushing myself away from the table.

"We seal the deal." And with that being said, he reached over, grabbed my neck and pulled me towards him kissing me and then roughly yanked my head to the side and bit my neck.

I screamed as I felt the sharp piercing pain of his teeth in my skin. What was he a vampire? Why would he do such a thing?! He did not take any blood though. Instead, he just left two large fang marks with two trails of blood oozing from them down the side of my neck.

"What did you do?!" I gasped, grabbing my neck in pain.

"I sealed our deal. Now when the time comes for our deal to be carried out, I will know where to find you, no matter where you are." Bryce said, and he snapped his fingers bringing us back to the dining room.

I felt dizzy and fell to my knees. Dawson ran to me and shook my shoulders.

"What has he done to you?!" He yelled.

He turned and ran towards Bryce.

"No Dawson, stop!" Mavus yelled, after he came and discovered the marks on my neck.

Dawson sent Bryce flying into the wall and at the same time I went flying into the wall. I hit my head with a sickening thud. I could barely breathe or keep my eyes open. Dawson looked at me, horror on his face. Bryce laughed getting back up onto his feet without as much as a scratch.

"What did you do to her?!" Dawson ordered him.

"Nothing, she just sealed the deal." Bryce explained.

"No. You liar! You marked her!" He held Bryce by the neck and I began felt a choking sensation in my throat.

"Dawson, stop it please!" I croaked, reaching for him, my head pounding.

Mavus was by my side, holding me. "Son you must stop at once! You are hurting her!"

"Let me go Dawson. Every time you hurt me, you hurt Trish." Bryce said, warning him.

Reluctantly, Dawson let him go and rushed to my side. Hejeena had gone to fetch my sisters. They came into the room just as Dawson got me up on my feet. Sherry, Tracy, Madison, and Rachel stared at Bryce in fear.

"I believe my work here is done." Bryce said, turning to leave the room, but then he stopped.

"Oh, before I forget. I forgot to mention one more thing before I let you fine people get back to your normal lives. I want you to meet my new second in command." Bryce said, proudly.

As if things could not get any worse, Luke walked into the door. Mavus walked forward with his staff, looking as if he were about to have a heart attack.

"Luke! What are you doing son? What have you done to yourself?" Mavus looked as if he would cry.

Luke had bleached his hair blonde and was dressed in a black robe similar to that of Bryce's.

"I have finally found my place father," Luke said and he smiled. "I have never felt so powerful in my life. Trish it is great, you will love it down there."

Luke smiled and his eyes turned black.

"My son has sold his soul to Bryce." Mavus spoke quietly, tears forming in his eyes.

Bryce and Luke left the room laughing.

"Is there any way to save him?" I asked, still holding Dawson, relieved they were gone.

"No." Mavus said, looking at me solemnly.

"His soul is damned. For eternity…" Hejeena whispered, shaking her head sadly.

"What do we do now?" I asked no one in particular.

"Well since you decided to make a childish

move, we have to figure out a way to stop Bryce from coming for you." Hejeena said, staring off into space.

"No I have to go to him, I made the deal." I said, boldly.

"No you will not you ignorant fool!" Hejeena argued.

"She is right Trish," Mavus said, "We have to get her prepared and we do not have much time."

"We have to find a way to destroy Bryce, before he starts this war. He is in the lead right now and he has my brother. I am starting to wonder if he was possessed." Dawson looked at Mavus.

"Well there is nothing we can do now. We have to fight them both." Mavus turned to him.

"I say we get started." Sherry said.

"Yeah, let us kick some ass," I said.

My sisters smiled at me, even Tracy and I knew that they were going to be there fighting by my side the whole way no matter what.

Chapter Ten

For the next five months I trained and did little of anything else. I tried countless times to contact Gabriella, but she never answered her phone, eventually it was disconnected. That scared me. I did not want to believe Bryce had kidnapped her and tortured her in the icy depths of the underworld, but I could not be too sure. Anger and revenge filled my mind, but I did not let the evil side of me take control. Instead, I learned how to make use of it.

Dawson, my sisters, and some of the rulers from the other planets came to visit and train me on different occasions. I learned, from Blaze, how to use solar power from the sun and turn kinetic energy into physical blasts that would hurt the enemy. He also helped me train with Draco, who learned a few fire tricks on his own. Draco and I learned to fly and combat in the air. I began to feel as if Draco was my child.

Against my wishes, Dawson brought Margaret

over, who showed me her telekinesis ability. Dawson was desperate, trying to get me to learn everything I could to defeat Bryce and if that meant bringing over psycho fiancé, then so be it. I picked the telekinesis up quite quickly, to her disappointment. She did not seem to understand that not only did I dislike her being Dawson's fiancé, but every time she used her powers, she used them against me to the core. As a result, Margaret and I ended up constantly using our telekinesis powers against each other. She made me trip over a branch that was never in front of me and I ended up making the server drop a whole pitcher of wine on her over a dinner Hejeena had held. Dawson finally had us stop meeting each other, because he said we were becoming "childish enemies". Whatever that meant…

By mid-July, I was able to use my powers in my favorite way. It irritated everyone else when I did it, but I could mimic others powers that I did not have myself. It took a lot of energy and it was only supposed to be used on rare occasions, or final attacks. Mavus told me that I should use it on a one-on-one encounter with Bryce, as a final blow. Bryce would not expect me, despite being the Chosen One and under twenty-one, thus not immortal yet, to be able to defeat him this early. Little did he know that I had another secret power, I healed rapidly faster than any other ruler. My progression with power and control had far exceeded anyone's expectations. I felt like Hercules in his prime.

August came around quickly and by that time I was a whole new person. I had successfully defeated each of my opponents in one-on-one battles. Hejeena and Mavus were both impressed. Hejeena actually stopped giving me such a hard time. Tracy was still a brat, but we were getting closer. Dawson and I spent most nights secretly together and our bond grew stronger. We were planning on coming out as a couple after the war was over, to keep the tension down and give him time to call off the engagement with

Margaret. I was trying to remain as optimistic as possible. All of us were, but we knew what was coming.

The only problem I was having was, specifically, with my special book. I had retrieved it back from Hejeena and over that past several months it had not shown me anything. It had not answered my questions or given me any hints on how to beat Bryce and prevent the war. I was beginning to get tired of the darn thing. If it did not want to show me anything anymore, what was its purpose? Then again, maybe the Father of Eternity was angry with me for some reason.

"Trish, hey, what are you doing?" Sherry asked, coming into my room and looking ready to train.

"Nothing at all," I said, putting down my book after my tenth attempt that morning to get it to respond to me.

"You are not ready to train yet. No offense, but ever since you have become superwoman, you have been waking me up." Sherry said, sarcastically.

"Yeah, I know. I just have a lot on my mind." I said, getting up and going to my vanity.

"I just bet you do. My goodness Trish, your hair has grown nearly too your knees! I love it!" Sherry exclaimed.

I gave her a funny look, "Are you feeling okay, because yesterday you said the same exact thing, except you told me that I should cut it off."

"Oh, well I changed my mind." Sherry replied, a little too quickly.

Something was not right, but before I could respond, Sherry had locked the bedroom door.

"Sherry, what are you doing?" I looked at the door and back at her, confused.

Sherry turned and stood in front of it, her eyes turning black.

"You are not Sherry are you?" I asked, suspiciously.

"Nope, sweetheart, I am just possessing her for

the moment and if I were you I would not try and hurt me. You see, if you hurt me you hurt Sherry and then I will just find someone else to possess."

"Who are you?" I glared at Sherry, despite knowing it was not her.

"Let us just say that Bryce has his friends that work for him in, I guess you could say high places or low." The demon laughed.

"Very funny, you are not supposed to be up here. This place is off limits to your kind." I warned.

"We have no limits Trish. Those boundaries these rulers up here have placed have already secretly been broken by us. I guess it is not a secret after all now that you know, but we are not obligated to the rules up here anymore. We only have one order and that is to bring chaos to all of the planets. Unfortunately, we are saving Earth for last." The demon sat down in the chair.

"Bryce has to keep his word, he made a deal. The war is not to start yet and it will not start yet, so help me"— I was cut off.

"Help you who? God, come on Trish. You think we are going to just sit back and relax while you build up your superpowers to defeat us? Not happening, Trish. Anyways, Bryce sent me here to warn you. It is not too late for you to surrender now and join him. He favors you a lot Trish and has big plans for you two, only if you give in." The demon suggested.

"Drop dead." I said.

"Funny, but okay, if you wish, but you cannot say I did not warn you. We are already up here and the universe is going to be a bloody paradise when we are through with it. The best part is that there is nothing any of you rulers can do, especially you Trish. You cannot save this universe. You are weak, alone, scared, and worst of all, you are still mortal." She said humorously, her black eyes shining and then Sherry's body suddenly started shaking violently. There was an electrical cold, vibrant feeling in the air and her body fell to the ground with a lifeless thud.

"Sherry!" I ran to her side.

She was terribly cold. I felt her pulse, it was there but it was faint.

"Someone help me please!" I yelled at the top of my lungs.

Sherry had most of her life drained from her body I was told by Claire, Neptune's ruler. Hejeena, Claire, and I sat together, no one touching their plates over dinner. It was almost as if Sherry was in a deep sleep or a coma. I felt like I could have prevented it and guilt hung over me. Dawson promised to stop by later in the evening. He and Mavus were calling all rulers to a meeting to prepare for war. Bryce had already broken the truce by allowing Sherry to be hurt.

"Look, Trish, do not feel bad about this," Claire said, her blue skin sparkling in the candlelight, "there is nothing you could have done. Most likely the demon that possessed Sherry is already dead because once a demon possesses a body, it loses its life and lives solely off of the body is possesses. It left Sherry, so unless it found another body within ten minutes it would perish forever. I think Bryce meant for this demon to die and wanted to scare us by hurting one of Hejeena's own. We rulers have great senses, we would know if a demon were to be present in our realms. He wants to get us anxious." Claire continued, looking at me, "If it would have stayed inside of Sherry, we would have had to perform some kind of ritual to release and kill the demon, or kill Sherry. This is serious dark magic Trish, demon possession is a serious power being used by Bryce. Bryce has already begun the war without declaration or our consent. It was an unwise move on Bryce's part, but he either has overestimated his superiors, or we have underestimated his powers. Either way, this is not good. We have to hold a meeting with all of the rulers and you of course." She looked at me, encouragingly.

"I do not understand, we made a deal," I said, quietly.

"Well maybe you should of thought about the deal you actually made! Bryce uses trickery. He is never to be trusted!" Hejeena snapped.

"Do not be harsh, Venus, she could not have known." Claire said her crystal blue eyes sympathetic.

"Claire, if you will excuse me and my daughter, we appreciate you coming to our assistance. I would like to be alone with Trish and I look forward to seeing you soon at this meeting." Hejeena said, courteously.

"Of course, you both have my sincerest regards. Good evening Hejeena." Claire stood, and went to open the portal.

Before she left, she turned and placed a long slender blue hand on my shoulder.

"Trish…" She said softly and that was all she said.

I looked back over my shoulder in time to see her close her eyes sadly and walk through the portal. After Claire left, Hejeena had the servants take away our untouched plates. Her hazel eyes looked terribly troubled. She turned those intense eyes to me and grabbed my hand. I instantly tensed because she had never done that before.

"Meet me in my room in exactly an hour. There is something I want to speak with you about." She said urgently, placing a hand on my cheek.

I just stared at her and finally said, "Okay."

My mother walked out of the dining hall leaving me there alone. I looked around at the guards, standing there like statues, and finally I could not bear the silence any longer, so I decided to check on Sherry. She did not deserve what happened to her. Of any of my sisters, she was the closest to me or at least understood me better. Rachel and Madison were distant most of the time and only trained with me when they had to. Tracy just liked to give me a hard time, but I knew that she would be the sister I looked to for guidance the most. I would have to attempt to have a talk with her, if that was even possible.

When I got to Sherry's room, she was lying in a black gown on her bed, her short hair combed back, sleeping soundly. Tracy was sitting in the chair next to Sherry's bed holding one of her hands. Her hair was pulled back into a tight bun and she wore a red sleeveless dress. When I walked in, she turned to look at me, her small brown eyes angry and then she turned back to Sherry. She did not want me to see the tears in her eyes. I felt terrible and I desperately wanted to make it all better.

"Tracy, look, I'm so sorry." I began.

"Do not dare apologize to me Trish," Tracy said, through gritted teeth. She got up from her seat and came to stand in front of me, her heels making her much taller than me and more intimidating.

"If it was not for you none of this would be happening! I thought maybe behind the shell you were strong enough to save us. Now I wish you would have just let Bryce taken you. You belong with him. Both of you are cowards, murderers, and completely alone!"

Tracy said angrily, her face red.

Before I could respond to her hurtful words, she hurried out of the room, the heels of her shoes stomping on the floorboards. I heard the echoes of her sobs down the hall, but all I could do was stand there in shock at what she had just told me. Tears did not even begin to cover what I felt like doing at that moment. I walked up to Sherry, lightly placing a kiss on her forehead and then turned to walk glumly back to my room.

I did not feel like meeting my mother. Everything was wrong and out of place. I walked depressed around the kingdom wondering what my mother could possibly want to tell me. Guards were everywhere keeping close tabs on me, I already knew. I pictured Dawson. He was probably fighting with Margaret right now. Trying to get out of their daily stroll to come see me and I knew it was not fair to her. She loved Dawson and even though he did not love her he had put himself in the situation to be with her forever… How was he to know I would stumble along?

When I got to my mother's room the door automatically opened and my mother stood on a balcony that overlooked the Moon. I wondered how I could even breathe in space. The whole gravity theory and everything I learned in school growing up seemed to be a childish thought now. With everything I was capable of doing now, that is. Hejeena wore a white evening gown and her hair blew swiftly in the wind. She did not even turn to look at me as I entered and walked up to her.

"Trish, I cannot in words begin to tell you how long I have waited to have you present in my life. The only daughter I had to give up and the pain I endured while you were away cannot be imagined. My heart disappeared the day I let your tiny seven pound body slip from my hands. There was never a day when I did not think about you and when we would be together again. It drove me mad. You have to excuse Tracy's bitterness. She was after all, my first child. She had it worse than the others because I could never give her the loving affection that a mother should provide for all of her children. Tracy and your sisters were cared for by servants, nannies, and of course, Mavus's sons. I anticipated your arrival and after I lost you my world shattered. Everyone wanted to give birth to The Chosen One. Yet, I swore when I gave birth to you, I wished it would have been someone else," she confessed, her eyes lost.

I did not know what to say. "You did what you could…mother."

She turned and faced me, a loving smile on her face.

"I was rough when I first met you and angry. I knew the family you had grown to love and I know they were taken away from you wrongly. I also knew that I could never replace the mother you had once had and that hurt me deeply. But as I think more and more about it, I would not have wanted you to have behaved any differently. You are strong, beautiful, brave, and even a little naïve, but in my heart I know that when the time comes for war, you will lead us all to victory and I can go take my place with The Father of Eternity," she said, and walked into her room going to her dresser.

I sat on her bed and watched her as she came back with a piece of paper that looked like it had been ripped from a book. She handed it to me. It was a picture of a sword. Not just any sword, but the most magnificent sword I had ever seen. The handle was encrusted with gems of all kinds, the sword gleamed pure gold, and had a bluish, red fire that descended from it. I looked at my mother with arched eyebrows. I had the chills.

"That book you possess, I know what it is because it was mine before yours. You see The Father of Eternity had all of this planned. He gave me specific orders and I left that book for you to find just as he had hoped. Now I have to tell you about this, now that you have mastered your powers. Well not completely, but good enough. This sword is called *The Sword of Good and Evil*. It was designed and created only to be used by the Chosen One in the War of The End. It hurts me deeply to say that the sword's whereabouts are unknown by anyone, but the Father of Eternity himself. His last request was for me to give you this paper and be the best supportive mother I could be. I know that you will have to go on a quest soon to retrieve this sword. Despite your mastered powers and being the Chosen One, this sword will kill anything or anyone that the Chosen One uses it on. Only the Chosen One can use it and if you retrieve this sword and kill Bryce, the war will be over. Bryce cannot know about this sword, no one can, because others will do whatever they can to stop you from possessing it. This is your destiny Trish, this is your fate. I know you can do it, but we must make haste because Bryce will come for you as your deal was made. He lies and he is powerful, dangerous, and deadly. If he gets you into the underworld then you will be trapped and everything will be on his terms." Hejeena was holding my hand tightly.

My eyes watered just a bit, but in my heart I understood. Unless I got that sword, the chances of defeating Bryce were slim. I had to make choices, smart ones, and I had to do something quick. The fate of the universe was on my shoulders and I was not even immortal yet. I looked at my mother and even though I saw her faith in me, I saw her doubt as well. Hell, I even doubted myself. How was I to find a sword when I did not even know where it was? How would I survive? I had never killed anyone before! What if Bryce found out about it? I am sure he had to of heard about it. Bryce would be here when I turned twenty to take me away. I had to depart within the next month! My mind felt like a rollercoaster.

"Excuse me."

I turned to see a guard at the door.

"Sir Dawson is here expecting Trish."

I looked at my mother and she nodded her head.

"Remember who you are Trish."

I nodded and walked quickly out of the room with the paper from the book in my hand. Dawson was standing at the bottom of the entrance steps as I made my way down the stairs to him. He looked at me, worriedly. I took a deep breath and went into his opened arms. I just wanted to be comforted. We embraced and then I looked up into his eyes.

"Dawson, I am going to have to leave this place in search of the only thing that will kill Bryce. I am going to have to leave soon." I explained, calmly hoping he would understand the urgency.

He nodded his head, as if he already knew this was going to happen. "I understand Trish. When must you depart?"

"Within the week, Bryce will be coming for me within the month and unless I leave now, my chances of escaping him are slim. If I embark on this journey he will have trouble finding me. We have to prepare for my departure. With what happened to Sherry I know that Bryce is not wasting any time starting this war. Demons are already swarming throughout these realms, I just know it, and I am not sure the rulers of the planets are prepared for what is about to take place." I said, walking towards my room.

He followed. "I agree. My father and I have been making preparations and have combined the military forces from all of the moons to come together for this battle. These demons should not be roaming around here freely. Bryce has unleashed them to impose fear and harm our kindred. I know about the sword Trish…."

"How do you know about the sword?" I asked, facing him, astonished.

"I have always known about the sword, from my dreams…I know that only you can find it and only with your hands will that sword have effect. No one else can use that sword, but you must find it before

Bryce does. I am certain he has been looking for it himself to destroy it. Unfortunately, I cannot go with you on this journey, but we will put together a trustworthy, loyal, and powerful team to accompany you." He said, sitting next to me on my bed.

I looked down at the piece of paper in my hand and then at Dawson.

"This sword's location is unknown Dawson. I am scared that I will not find it in time. It could be anywhere, and I have not even been to any of these planets or galaxies. I am just a human, ordinary girl that got sucked into this fantasy world. As much as I want to give up, I know that it is not an option. Surprisingly, I have grown to love my family, Draco, and you. I feel strength inside of me that I never knew I had before. Sherry is in coma because of me and I will not allow anyone else to be hurt. I need you to sit down with me tomorrow and help me map out a quest Dawson."

He nodded in agreement, "I will assist you with

everything that I can. I am not going to lose you Trish. If this universe falls apart I am going to go down with it. My brother has betrayed his own bloodline. Never in history has any heirs or heiresses of any planet betrayed their own kind. Bryce has taken this too far and whatever he is planning to do is going to be massively destructive. My greatest fear is for him to capture you Trish. His father, Xandran, is the ruler of the Underworld. He is the most horrifying, gruesome, being to ever exist. If anything happens to you and somehow you get captured, I am going to die saving you."

I looked into his eyes at that moment and my heart froze. I was truly in love with Dawson. His deep blue eyes were so serious. I leaned over and kissed him with everything I had. He pulled me into him kissing me hard, more fiercely. I lost all sense of reality and I felt my adrenaline soar. I knew I had to pull away from the kiss because our powers were beginning to radiate off of each other. I knew he felt it too and he unwillingly pushed me gently away.

"I have to go. We will meet tomorrow morning. I am going to put together a solid team of five of the best comrades I can find for you. Do not worry about anything except packing your necessary travel items. I will have Griffin arrange a space shuttle for you and your team, however, you will still need to learn how to open and close portals between the planets and other dimensions. It is a dangerous universe out there Trish and we must have means of communication, via the tele-communicators that I and my military personnel use. I will show you how to use those tomorrow as well. Draco will be going with you, but he will be flying with the other dragons. They will be with you on most of the planets you land on, but Draco and the other dragons cannot accompany you to Earth, the planet inhabited by mankind."

I shook my head, acknowledging everything he was saying. "You are a professional Dawson." I said, lightheartedly.

He stood, taking my hand into his. "No, my dear I am a King, I am a son, and you are my future. I will do whatever it takes to protect and honor all of these reputations."

I did not respond and he touched my chin lightly. "Until tomorrow," he said.

"Until tomorrow," I replied, my heart sunk as I watched Dawson leave my room.

After Dawson left, I walked through the vacant halls of the kingdom in search of Tracy. I wanted to talk to her, to reason with her, and to try to make things right between us before I left. In my heart I knew that I would need her help. I found her in the sitting room having tea with Rachel and Madison. When I entered the room they all looked at me and silenced. Tracy gave me the onceover and looked away in disgust. She was such a snob! Rachel and Madison looked back and forth from Tracy and me and then looked down at the floor. Well, if they wanted to give me the cold shoulder fine, but I needed my sisters on my side, especially Tracy.

"Rachel, Madison, could you give me a minute alone with Tracy please," I said, sweetly.

They looked at Tracy unsure, like little pets, with a questioning look. Tracy groaned and then like the princess she was waved them to go. "Let us talk."

Rachel and Madison got up and left the room whispering amongst themselves. I took my seat in one of the empty chairs facing Tracy. Even angry Tracy was remarkably beautiful. She was dressed in a fitted black gown and her hair was done in a voluminous bun on top of her head. She glowered at me with her small dark eyes waiting for me to speak. I felt belittled sitting next her in my plain white t-shirt, jeans, and sneakers. My hair was an unruly mess of tangled curls and I had not even bothered doing my makeup. I was tired, grumpy, and ready to get this fight over with.

"Tracy, I need you to accept my apology about Sherry. I do not want us to be enemies anymore, you are my sister. I know you are aware that I am going to be leaving and as happy you are about it you have to understand, really understand that the fate of the universe rests in my hands. I am much younger than you probably by centuries. Your resentment is not something that I need right now rather it is your partnership that I need to fulfill my duty. I had a long talk with mother earlier and I have to go on a quest to retrieve an important weapon to kill Bryce. I am scared Tracy and I am doubtful of myself. When I look at you I see your pain, your bravery, and your power. I know that you are familiar with the planets, know the rulers, their laws and their territories. I need you to help me find this sword. Most of all, I need you to believe in me and want to save this universe with me. If I fail Tracy, the universe and everyone in it, including you will die. I know you want so much to see me fail, but trust me you do not want to see me fail at saving the universe."

For a moment, her face softened and then she looked away from me, staring off into space. I could have sworn that she had stopped breathing and forgot that I was even there.

"Tracy, please." I pleaded.

Snapping out of her thoughts, she turned and faced me.

"You know something Trish? I used to try to please mother in every way a daughter could for her mother. I worked hard for years just to get her to praise me. She was always so determined, lost, and focused on something else. Something that did not even exist! I was her first daughter and her most loyal servant. I did everything right, but it was never enough for her. She was waiting for someone, another daughter. She would talk about you as if you were real then and there! I could not stand it and when I looked in the mirror I hated myself. I hated every single thing about me." Tracy said, icily, tears forming at the corners of her eyes.

"Tracy, look I," I began.

"Let me finish…" she said, coolly.

I sat back in my chair and hushed.

"Then the other girls came one after one after me. She gave each of us a certain level of respect, more motherly to Sherry and Rachel. Madison and I are of tougher skin. I earned my respect as the dominating daughter from the other girls. They abided by my rules and my calls. Instead of drowning in my sorrow of winning a mother's affection, I got all of her daughters to look up to me as mother. I pretty much stole mother's title. I took over a lot of the queen's roles. I travelled the planets, made diplomatic trips with rulers of different galaxies and planets, and I was looked upon by the community as a queen in her own right." Tracy said, her tone stronger more bold.

I looked at her, sympathetically, but oblivious of any emotion. She was not going to make me feel guilty for being born. She continued, "Everything was under my control that was until you were born. When you were born it was as if the universe froze for a very long time. Mother took initiative the second she realized she was pregnant with you. It was ironic because mother's pregnancy was revealed out of nowhere and the mystery of who your real father is… No one knows, not for centuries. She put Mavus through the worst just to have your ensured protection forever. Everyone knew about the Chosen One, but mother did not want anyone to know about you. She would have dreams, panics, visions, and sometimes lock herself up in her tower for hours of the day, sobbing. It scared us. We did not understand what was going on. The Queen of Venus was having a breakdown and us, her daughters had no idea who you were, but we despised you. Everything mother did was for you. She devoted her life to you, your protection, and success. Your sisters, Trish, had to devote their lives as slaves to some unknown sister and when you arrived, I was furious. You were the most flawed, strange, and unimpressive human being I had ever seen…."

I was getting slightly irritated. This was one

heck of a confession, but I still wanted to push to the point. It was nice to get some background information on my family history and I continued to learn the reasons for the resentment I received from my sisters, but I understood it enough and it did not bother me. Still, I had the fate of something much greater than sibling rivalry on my plate, so I hoped Tracy would finish her point soon. When I did not say anything she continued on.

"Regardless, I adapted to you being here and I have come to terms with your purpose. I have realized the error in my ways. When I have been against you I should have been assisting you. Your role is more critical than I had anticipated. I want to apologize, Trish. I have been such an angry woman when I should have been preparing for this battle. You have a heavy weight on your shoulders. You are but a child. I feel that you have by far the worse fate than any other person in existence. But we must persevere… You are still my younger sister. I have hundreds of files, maps, and manuscripts that might prove helpful to you on your journey. I want you to come to my room later on this evening so that I have time to get them to you. There have been many ancient legends, stories, and quests that have been carried out all over the universe. Man only knows so little but there are passageways, dimensions, and worlds that the average mind cannot fully conceive."

I was at the edge of my seat. Now she was talking! I needed to see these papers or whatever they were. By this point in the conversation Tracy's tears had dried leaving two trails of dried mascara down her cheeks. I stood holding my breath and went and took her into my arms. She was taken aback and tensed under my touch, but I did not care. I hugged her with all I had and her shoulders loosened. Tracy slowly hugged me back. I was so relieved.

"I will meet you in your room later on this evening." I said, letting her go.

"Do not think that this conversation alters in

any way our relationship Trish. If anything this is just a mutual obligation between two sisters that are bound to the same fate if they do not cooperate." She said, gaining back her self-composure.

I smiled and said, "Understood."

Chapter Eleven

It was always hard to tell the time in space because in actuality there was none, but I knew it was getting closer into the evening hours for the kingdom was beginning to get darker. Hejeena liked her castle to be dim and there were only candles hung along the hallways providing a faint light for whoever was wandering through them. I knew I was supposed to meet Tracy later, my mother was arranging a grand meeting with all of the rulers before I left for my quest and Dawson was making all other preparations. I felt helpless at that moment knowing there was plenty I could be doing to prepare, but my mind was completely out of it. I had grown accustomed to a life of routine, of practice, and power. I had forgotten what life on Earth had even felt like. My family, Gabriella, and my other life experiences felt like a distant memory now. My anxiety had gotten worse and I could not seem to concentrate.

I decided to explore parts of the castle that I

never had before, maybe pay a visit to Draco because I knew in my heart I may never be coming back to this place. I walked the vacant stone hallways. My steps were soft and I could feel the coldness of the marble floor with my bare feet. For anyone else walking alone throughout this castle it would appear to be frightening, but I knew that I was okay. I would turn multiple corners not opening any of the closed doors not memorizing my way back, but just going with the flow. Hejeena had multiple plants and statues across the whole kingdom to provide for some type of dull decoration. I was trying to formulate a plan in my mind on how to retrieve the sword when I saw a faint red glow coming from one of the rooms at the end of the hallway I was walking through. The door at the end was slightly ajar and uneasiness crept up in my stomach.

I made my way closer moving slowly not letting my curiosity get the best of me. All of the doors had been closed for this long, so why would there be this one open all of the sudden? It seemed rather odd to me, but I was entranced by the red light coming from the door. The closer I got to the door the more sudden pain I began to feel in my neck. I remembered Bryce's marking on my neck and froze. Why was it that every time I moved closer to the door my neck hurt? Was Bryce in there? No it could not be because when Bryce felt pain then I felt pain too. Bryce did not want me to go into that room. I would take it though because I wanted Bryce to feel pain whether I felt it too or not. I continued to walk closer the pain in my neck growing more and more with every step I took. I grabbed a hold of the wall wincing. I could do it! I had to find out what was in that room!

By the time I got to the door my nose and neck were bleeding. I pushed the door open with everything that I had. Inside of the room the red light blinded me and I was sucked into the room, the door slamming shut behind me. The light was too bright, the pain in my neck too strong, and I could not breathe. I fell to my knees, covering my face. The red light was blaring and overbearing. I heard a booming voice from nowhere.

"My dear Trish, how is it that the more and more help I give you the more disappointed I become? You have acted foolishly my child. You have bound yourself to the Devil! I should not have placed so much on your plate at once, but you are the only one who can carry out this mission child." The voice said, loudly.

I opened my eyes, but the light was still too bright and I could not make out anything in the room.

"I am so sorry! I do not know what you expect of me!" I cried to the voice.

"Little Trish, you do know what I expect of you. Open your eyes do not be afraid of your destiny! You know about the sword now Trish and you know that you are meant to kill Bryce. Ask yourself, how far are you willing to go for life? Can you imagine? Can you really imagine the universe Trish and all of its components? This is not just a duty, my dear, but a privilege! Life is the most amazing gift ever created. The ability to love, to feel, to experience, to sacrifice…those humans have it good down there and despite their ignorance and foolish ways they are your primary responsibility to protects. I have watched the world transform before my eyes since the beginning of time and I knew that this day was going to happen before it happened. Trish, I am your father, Bryce's father, everyone and everything's father. Yet, for some reason I have lately been caught in a cloudy storm. I have found myself sad and angry at the conduct of mankind, the planetary rulers, and of you Trish. I have chosen you to fulfill the impossible, the unexplainable, and to conduct the ultimate sacrifice. I know you will not let me down, but you must not lose your faith. Now that you are bound to Bryce you have connected yourself to the forces of evil. Bryce will be able to find you sooner or later and you must be prepared for this now that you are bound to him."

I felt the tears sliding down my face and I did not know what to do anymore.

"I am a failure Father. I have been a failure my whole life. I have lost everything and everyone I have loved. I am a walking curse! I just want it to all be over!" I screamed.

"Silence now child! There is no time for self-pity or self-loathing! Raise yourself above the darkness, Trish! I am going to help you for the book at this point will no longer be needed on your quest, but keep it close by your side for little hints will show up on its pages along the way. That is how I will help you. Now listen to me carefully, every ruler has a birthstone similar to the one that you have that your mother gave to you or like the one Dawson gave to you. Nevertheless, each ruler keeps their birthstones in sacred places where only they know their whereabouts. In order to retrieve the sword you must meet with every ruler and collect these stones. Only after you have each of the stones will you be able to open the case that contains the sword, which is located somewhere on Earth. You will know when the time comes. The only downside to this is that Xandran, being a ruler himself, has a stone that you must retrieve as well. Unfortunately, getting his stone will be the hardest, but I have the utmost faith in you Trish…This is all I can help you with for now, but be brave and do not lose a sense of your true self. For if you lose yourself you lose everything." His voice was slowly fading away.

"Wait, do not go! I need your help! I cannot possible find all of these stones in due time! I need more than this Father!" I cried.

I heard a loud chuckle, "Ignorant child, there is no such thing as time!"

Suddenly there was a loud bang and I covered my ears with my hands crouching down. When I looked up the room was dark and there was complete silence. I slowly pushed myself up off of the ground. My ears were ringing, my neck hurt terribly, and I had accidently wiped a smear of blood across my face from my nose. I felt lightheaded and exhausted as I made my way out of the room. At this point, I wish I had remembered the way I had come. I decided to continue on down the next hallway because I heard Draco's roar in the near distance. Oh, Draco! My beloved dragon, he could sense my pain. I stumbled down the hallway and out a large double set of doors onto the training field and dragon stables.

Draco was flying around the training field furiously. The guards and dragon keepers were trying to calm him down. The other dragons were lounging calm and lazy. It amazed me how enormous the dragons were, but my Draco stood out the most. He was the largest, fastest, and the only blue dragon there. The guards were relieved to see me, but when they saw the dried blood on my face and the redness in my eyes, they were concerned and they all stopped worrying about Draco to come to me. But I was not paying much attention to them as I stumbled across the field my arms reaching up to Draco. He could not see me though. He was flying around in circles like a mad dragon.

“Draco! Come to me!” I screamed at the top of my lungs.

At my call, Draco instantly stopped flying and when he saw me standing there, the tiny dot that I was, he roared. It was not a fearful roar, but a protective one and he did a U-turn coming down with a loud thud just feet away from me.

“Trish, are you alright?! I was terribly worried about you! I sensed something was wrong.” Draco spoke, his voice hoarse.

I went up and hugged Draco, though I know that he probably only felt the slightest sensation from it.

“Oh Draco, it seems just like yesterday when you were just a baby.” I said, climbing up on top of him.

I had never seen a dragon roll its eyes, but

Draco did. "You are dodging the question Trish and why is there blood on your face? My goodness, what has happened to you?"

Still not responding, I rubbed my hands along his shimmering blue scales closing my eyes. He felt so rough, but it was comforting. "I love you Draco. We are going to get through this right?"

Draco grumbled. "Your disregard for my concern is rather discomforting, but I love you as well my Trish."

"Let us fly, we can talk later. I am tired Draco." I began to doze off.

"You are not the only tired one, but sleep now for you will not be getting much sleep in your near days." And with that he took off. I felt the wind blowing my hair rapidly as we flew around the field, but it felt great and I knew I was safe with Draco. Here I was trying to save the universe and taking a nap ride on a dragon's back.

The nap only lasted about thirty minutes, but it

was the best nap I had ever had. I remember slightly opening my eyes and seeing myself flying through the clouds. They looked different than they did from looking up from on the ground. I felt Draco's muscle movements, he was at ease flying, I could tell. He swooped down so suddenly that it woke me up. I looked down into the training field to see all the guards standing there with Hejeena, my mother, in all of her glory and superiority. She had her hands on her hips, long curly hair let down, and she was wearing a red gown. Clearly she was upset and impatiently waiting for me to come to her.

As if reading my mind, Draco said, "Well, are you prepared to be handed over to the hands of suffering?"

I chuckled and let out a sigh, "As ready as can be."

"You do understand that the first thing she will do is go into hysterics about that blood on your face."

I had completely forgotten about it. Draco

landed a considerable walking distance from my mother. I jumped off of him landing on my two feet, but a jolt of pain shot up my legs. I had landed too hard.

"Take it easy Trish, do not overdo it," Draco said, caringly.

I smiled. "I think I can handle this and thanks for everything Draco. It is crazy how fast you have grown. I remember when you were this tiny baby, blue dragon that…"

"Trish! Get over here this instant!" I heard my mother's voice, as if she was standing right next to me, but she was a few yards away. I cringed.

"Go on now. I got to go train with the other dragons."

I nodded, patting him on his side and then headed over to face my mother's temper.

When I walked up to my mother her eyes widened, but she did not look surprised.

"Tell me, your face?"

I looked at the ground, but then I met her concerned expression, there was no time for lies.

"He came to me, Father did. He is disappointed in me, but I must collect the stones…."

"Hush! Trish, do not speak of what the sacred tongue speaks to you. It is meant for your ears alone. Otherwise, what you speaketh will be heard by unseen ears." She motioned for me to follow her.

I obediently kept quiet as we walked quickly through the halls of the kingdom. I could tell she was anxious. She kept glancing back at me and I wondered what was wrong. I knew I looked horrible, but what could I do? I hated how vulnerable I felt when everything was on my shoulders. I had profound doubt in myself. We went all the way back into the throne area and up the stairs to my mother's chambers. Why was she not speaking to me? Something was up. Hejeena walked to her closet and pulled out a magnificent silver dress and laid it down on the bed. I could not help but touch it and feel the sparkling silk fabric underneath my fingers.

"You will be wearing this in approximately an hour. There is no time to waste. I did not want to stir any unnecessary emotions from you. However, all of the rulers will be here for dinner tonight. You must speak to all of them. I know you are nervous, but we must do what we will before matters get worse. Word across the planets is that Bryce's demons had broken through the protective shields of nearly all of the planets by now. It seems that he is moving in slowly, but establishing unidentified forts on other ruler's premises. Civilians of some of these planets have already gone missing, some possessed, and others losing control of their powers. I need you to speak to the rulers individually and as a group. You know what you must do and what Father has commanded of you. Now go and wash up, use my facilities. Prepare yourself and whatever you do, do not show fear. Be strong, bold, and courageous. This is going to be nothing that you have ever experienced before."

Oh I could believe that!

"I do not think I can do this. I'm on the verge of being twenty years old. I am about to go on a quest that could take me ages to complete. They will look down on me, mother… I highly doubt these rulers will give me their birthstones and trust that I will accomplish what I have to. Apparently, this box is located somewhere on Earth, and that is all I know. I cannot possibly go into the room and explain this too them!"

Hejeena placed her arms on her hips. "You are so predictable, Trish. Do not make it a habit. Your enemies can read between your lines really well. No wonder Bryce wants to make you his Queen. There is no time for discussion and pleas. It is a fact of the matter situation. Every one of those rulers will comply with the order you, The Chosen One has issued. Do not forget your superiority, Trish. Though young and courteous you are in control of everything and everyone's lives here. Do not allow centuries of outdated rulers to discourage you. No, my dear you are the remainder of a long battle. You are the final contender. Do not fail." Hejeena came and kissed the top of my forehead lightly.

She turned and left the room going to one of her towers probably. To cry, I thought. But her words touched me and I knew she was right. I went into her bathroom to get cleaned up. My mother's bathroom was quite simple and plain compared to the rest of her luxurious living surroundings. The room consisted of dull stone walls and flooring. It was ridiculously cold in the room and a gargoyle statue was perched on the ceiling with warm steaming water coming from its opened mouth. To be honest, it kind of gave me the creeps. Large red candles were lit all over the bathroom, but that did not keep me from shivering as I undressed and slowly walked to stand underneath the already pouring water.

The temperature of the water sent a jolt of electricity through my body. This was not the average type of water I was used to. Everything about the water was uplifting. I instantly felt better and closed my eyes, determination growing. I could do this. They were probably arriving by now, getting set up in the dining hall. I wondered how my mother had set up the room. Not all of the rulers had the opportunity to meet me, especially after that embarrassing moment with Luke at the ball. It seemed ages ago, but I knew that my first impression to the rulers had not been a good one. In my heart I felt that they despised me. After all, I was just this ordinary mortal child that was blessed and cursed with the fate of the universe in her hands. They could not possibly understand how I felt about that. If I could have given the title of the Chosen One away I would have in a heartbeat. But it was not going to happen. It did not take long for me to get ready and to put on the silver dress that my mother had given me. It fit perfectly and hit the floor leaving a slight trail of fabric behind me. I entered back into my mother's room to see her standing there dressed in a red sequined gown, ruby pins holding up her curls. She was holding a large gold box in her hands smiling at me. For a moment, I felt she looked proud of me, but then her dark eyes glazed over and a strict expression crossed her face.

"A true queen must wear her crown. As the Chosen One and the future Queen of Venus, you must not only wear this crown but walk into this room it being your first and may very well be your last conveyance of a crucial meeting among the rulers of the planets. I would like to present you with mine and my ancestor's crown, Trish."

My heart was beating rapidly. I looked at her nervously as she sat the box down on the dresser. She lifted the lid, glanced at me, and pulled out the most beautiful crown I had ever seen. It was a solid gold crown with silver sparkles embedded all over it. There were diamonds and rubies aligned along the circlet, and in the center of the main point of the crown was the largest ruby I had ever seen. She softly smiled at my fascination with the crown. I could not take my eyes off of it! Hejeena stood before me and slowly put the crown on top of my head. It felt heavy but secure. She took a step back looking me over. Satisfied, she said, "You look like a true queen."

I believed her. "Thank you."

"Now let us hurry. A true ruler never wastes time."

She did not waste another second taking off down the hallway. Suddenly, I was uneasy. I took a deep breath and took off after her. She stopped just before we reached the entrance of the dining hall. She turned and gave me a straightforward look.

"Trish,' she said, 'good luck. Make me proud."

I did not respond but clenched my jaw. Purpose took over my demeanor. I walked into the room and all of the bickering that had been going on instantly stopped. A dreadful silence hung over the room. I took this as my opportunity to observe my surroundings. The rulers all sat amongst each other at the long oak table that was filled with mouthwatering dishes and various deserts. There was no time to have an appetite. I wanted to look each of the rulers in their faces. There were quite a few and I needed all of their trust and cooperation. They all stared at me, some with curiosity, others with doubt, some with pity, and others with envy. Each one of them had a game face on, their defenses up. My mother took her place standing behind me. She lightly patted my back. I cleared my throat.

"Introduce yourselves." I ordered.

They each looked at each other questionably and my courage slightly faltered. I looked back behind my shoulder at my mother. She nodded.

Again, this time a little bit louder, I said, "That was an order!"

The tone of my voice released almost a shockwave of my energy across the room. It only released a soft static, but it was enough to make the ruler at the end of the table rise up. The ruler that stood first was a female. She was tall, awfully tan, and thin.

She had absolutely no womanly curves. Her facial features, however, were exquisite and her tanned skin was flawless. This woman's orange eyes were bright and slanted, catlike. Her cheekbones were high, small red lips pursed into a thin line. She wore a one-shouldered sand colored gown revealing the hundreds of brown freckles sprawled over her skin. Her hair was a vibrant orange-red that was cropped into a pixie-cut around her face. She looked to be in her late twenties, but as always she was probably ancient like the others. I waited for her to speak.

"I am Miranda, the Queen of Mercury." Her voice was deep, but clear.

I nodded and bowed my head with respect. She remained standing looking as if she wanted to test me. Her gaze was secured with mine. My body was suddenly warmed and I felt a slight increase in the temperature of the room. I kept looking at her knowing she was the cause of this.

"My power is fire, heat, and temperature

control. My planet is next to the Sun and closest to it. I serve as a messenger between the Sun and Venus. As my duty, I will help in whatever way I can."

"Thank you Miranda."

She sat down. Blaze stood next. I recognized him immediately remembering the moment he gave me Draco. Sonya stood with him. I smiled at them. He looked as handsome as he was from the first day I had met him. Sonya was the same too. She stood without speaking behind her husband, her attire not being much different from Miranda's, just a couple of shades darker. Blaze was wearing a velvet robe embroidered in fine black designs. He had a concerned, but peaceful expression on his face.

"Queen Trish, it is a pleasure to be in your presence once again. My wife and I understand the universal state of emergency we are in. We have come fully aware of the sacrifice needed and wish to assist you in anything necessary."

I smiled. "I appreciate you both, greatly."

They both took their seats. I quickly looked at the next ruler. We had no time to waste! I needed to know their names and where their stones were before Bryce came for me. Next a man stood. Yet, he was not the average looking man. He was excessively muscular his head bald, yet a long unruly beard went down to about his midsection. He looked at me with red angry eyes and scoffed as he rose. He wore no shirt, no shoes, and his leather pants were way too small on him. He was a fearful looking man, his height towering above most of the other rulers. I tried to keep eye contact with a straight face.

"You are?" I asked, softly.

"Graydon, I govern over Mars. I do not have time or the patience to put my faith in a little girl, while my planet and my kingdom is in jeopardy. I have already experienced enough invasions from mankind. I need to allow my armies to destroy whomever Bryce sends my way on my own. I cannot wait for your call." Graydon stated, angrily, looking at the rulers trying to get them to see his side. They all looked away from him.

"Graydon, we do not have time for your anger and we do not have time to dispute the best way to end this war. I need to know about each of you and I need to obtain crucial information from all of the rulers, privately, to make sure that just in case one ruler decides to betray us then they will have a difficult time helping Bryce. We are not here today to enjoy each other's company and be allies. We are here because each of our planets, our kingdoms, our families, friends, and the entire universe is on the brink of destruction. What Bryce is planning is monstrous and will cover the universe in nothing but pure blackness. He does not care what he does at this point except to bring chaos. So please, Graydon, my life is invaluable at this point compared to the ruler's ancient bloodlines. We need to proceed and as for anyone else with personal opinions and feelings lose them right now. I do not want to hear it." I said, strictly daring anyone to test me.

I thought I had gotten my point across when the floors began to rumble and I could see Graydon staring at me, fuming. Enough was enough, what would it take for him to learn? I looked at my mother, she would not speak for my protection and it would make me appear weak. Graydon was testing my authority and he need not try. I closed my eyes facing Graydon. How could I make this as simple and warning as possible? When I opened my eyes they were pure silver. They penetrated through Graydon's angry red eyes and I was able to sense the tiny flame of fear that he had all burrowed deep inside. I saw his immortal heart and in my mind I clamped my hand around it, squeezing his tightly. Unsure if it would work, my invisible mind power, Graydon clutched his chest and his eyes widened as if he was having a heart attack.

"Sit down, Graydon. There is no time for earthquakes and there is no time for Mars to be without a ruler right now, but cross me again and I will blow you into pieces from here."

Graydon sat down fast, in disbelief. The rulers seemed to be afraid of me now. I did not want them to fear me, but sometimes it is the fear that will keep them by your side. I knew my evil side was enjoying this, bottling up their fears and feeding off of them. I did not want to see Dawson's face right now. I was certain he was appalled by my actions. Claire stood next and I was calm once again. I knew she was a close friend of my mother's.

"Trish, I am Claire, the queen of Neptune. It is a pleasure to accompany you and to assist you and everyone else in this war of the unknown. Let us not battle each other, we are not the enemy. Bryce is… Xandran, his father, is. We must stand strong, deliver all we have been given. Our power, our strength, and our courage, and dismiss the negative prejudices that you all carry of Trish. The fact of the matter is that it is her birthright to be the Chosen One and she must fulfill her tasks. Given her age and her mortality still, it is our duty to help and protect her. Now let us do this without a fuss." Her blue eyes held such authority.

I saw a few of the rulers nod in agreement. Claire's blue skin sparkled and the white glittering gown she wore gave her a majestic appearance.

"Thank you Claire. Your cooperation and words are dear to my heart."

Claire sat down and another ruler stood. I felt like these introductions would never end and I was already tired of standing with a heavy crown on my head for such a long period of time. The lady that stood up next had long white hair, gray eyes, and was very petite. She looked like a frail old ghost. Her skin was deathly pale almost appearing to have no color at all and she wore what a gray wool tunic.

"Greetings Chosen One. They call me Arayan. I am the queen of Saturn, the ruler of the magnificent rings. Since I reside a great distance away I am closely affiliated with Claire of Neptune and Pradon of Pluto. I know that since our kingdoms reside quite a ways away we will all assist the other planets and your needs first. Given that our planets are composed of gas and our kingdoms hidden, Bryce's threat to us will take time. Therefore, I salute you and your purpose, Chosen One. I will assist you as any servant would to his master. It has been a long while since I have had such an honor."

That earned a smile from me. I knew that I could trust Arayan, Claire, and Pradon. Pradon stood next, Arayan took her seat. He was the same as I had last seen him. His purple skin, bald head, white eyes, and sharp teeth were the same and he still wore the same black cloak he had worn at our first meeting.

"Arayan speaks for me as well. I am Pradon, the king of Pluto at your service." He rasped.

I nodded, respectively.

He sat and I quickly looked to the next ruler. We were getting closer to being finished. A man of athletic build stood, yet he was perhaps the most interesting to look at yet. His skin was silver but it glistened with radiance as if it was extremely moist or even slimy. He wore a light gray robe that blended in with his skin. His eyes were bright artic blue and his lips purplish blue. I suddenly felt the coldest I ever had felt in my life and had to close my eyes to keep my teeth from chattering. A cold wind began to circulate within the room and I shivered.

"I am Lupi, ruler of the planet Uranus. Never have I been in the position to bend to the will of a rather young lady, but what has to be has to be and what has been foretold will be done. Let me provide my services from Uranus and hope that victory will be ours." He said, coldly without emotion.

"Thank you Lupi."

He sat down running his long fingers through his silver short hair. A dark man stood next, his skin the color of soil and his eyes the color of a horrific storm. He was very handsome, but he seemed to be on edge. He was dressed in an earthy colored tunic and he had orange shaved hair. I waited.

"Krypton of Jupiter and here I stand to offer my companionship. My kingdom has always remained safe even from mankind thanks to the massive storm circulation that Jupiter provides. It unnerves me what Bryce may be planning for an invasion of my planet. My nights have been sleepless, my community shaken, and my storms outrageous. Jupiter is by far the largest of all of the planets and I fear that Bryce will seek to destroy and conquer it first as a means to help him win."

"Do not fear Krypton. That is what he wants. Jupiter will be safe for its storms provide sufficient security for now. If anything, Jupiter should be last on Bryce's list because of its huge mass and chaotic atmosphere. Let us continue on and make haste though. We can never underestimate our enemies." I said, reassuringly, surprised by my own wisdom.

Krypton did not say anything, but instead he sat down, still he looked nervous. I wanted to tell him it would be okay, but I did not want to make any promises that big, because truthfully things might not go smoothly and I did not want to allow my doubt to discourage them. Mavus, Noreen, and Dawson all sat towards the end of the table and I knew that I would not have to have them speak. Dawson and I exchanged glances and I looked at the last ruler, besides my mother, sitting at the end of the table.

He looked the most human of us all, well, besides me. He had Asian features and blonde spiked hair. His jeans were faded, he was barefoot, and wore a slightly unbuttoned white collared shirt.

"Well, it all boils down to me,' he said, standing, his green eyes staring into mine, 'I am Jiles, ruler of the Sky and co-ruler of the Earth."

I could not help but blush as he looked at me for his features were purely angelic.

"As you can see my brother is not here, for it is he that is why we are here. I apologize for what is taking place, but if anyone understands Xandran it is I. We govern the same planet, Earth. I have devoted my entire life to the humans. I have watched their rise, their fall, their pain, struggles, births, deaths, relationships, temptations, and needs. Humans are lost, easily manipulated and fooled. I have seen Xandran conquer and control many, millions even and there was nothing that I could do about it. Humans are the most important creation and our purpose depends on protecting them as much if not more than our own kingdoms. These creatures are doomed for their selfish ways and impossible to satisfy anymore. The days of glory, nature, love, peace, and harmony are over. I have resorted to the sky and constantly battle the environmental damages and pollution to the atmosphere. We have to understand what we are up against. Xandran is so powerful primarily because he has the capability to control, persuade, tempt, and delusion the human species. Oh, it depresses me so! They all fall into his hands like ants to sugar or bees to honey. I fear for the worst! He has been unleashing demons from his under realm for centuries. They walk the Earth, limit my power, and destroy. Earth is the one planet we must save and protect before it is too late. Xandran and Bryce do not really need to conquer our kingdoms, all he needs is to fully control the Earth and destroy us. I watch from the clouds and try to send signs, for what do humans truly understand about death and their souls? They live consumed in their day to day lives ignoring the possibility of the true future, though they plan and invent they destroy more than they can save thanks to Xandran and his wicked ways. I can only hope that you Trish can save us all."

Jiles speech had brought tears to my eyes, but also a surge of hope and strength. I stepped forward facing him primarily.

"Let me tell you something about being human, all of you,' I faced all of them, 'us humans may very well be weak and ignorant, but we have the ability to feel and experience the life of which you ancient beings will never fully grasp or understand. Can you imagine how it feels to be so small and alone in a world of billions?! Could you even for a second grasp the feeling of being powerless, but still fighting, still having a heart the greatest token of them all, to hope, to pray, and to continue on? It is not easy for us down there and it never will be. We do not know any better, but to go for our desires. How can we live in peace and harmony when our ancestors and their ancestors have already dug our grave so deep that all our future generations can do is lie down in it and continue their unfinished products? There are, though, the ones that still keep purpose to our Earth and are the reasons we must defeat Bryce and never give up hope. Those are the children, the innocent, the poor, the struggling and the tired, the elderly, and the sick. Oh, and do not forget the humanitarians, the dreamers, the teachers, the artists, the lovers, and the families. Jiles, remember your position, your potential. If I remember correctly, you have just as much power, if not more than Xandran. The good always conquers the bad, always."

He did not say anything to me and I did not

expect him to. I could tell the rulers were beginning to get restless. I had finished the introductions. Now I had to retrieve the stones.

"Before I dismiss all of you, I need each of you to meet me in private, briefly. These private meetings are to be confidential between me and each ruler alone. For information leaked may lead to an early defeat of us and destruction of all of our kingdoms. All it takes is one to betray us and we all fall." I said, gravely.

I left the room quietly and got situated in one of the isolated chambers. The room was small, one chair sat in the room and candles were lit everywhere. I had each of the rulers one by one, at Hejeena's command, come into the chamber, sit before me and write without speaking, each on a separate sheet of paper from my sacred book, where I could find their birthstones. I knew this was the safest and fastest way I could think of without blowing anyone's cover. I knew that everyone had special powers and I did not want to risk anything being made known to Bryce that could knock me off of my path to the sword. I knew that Father protected this book.

Dawson and my mother watched over the rulers while, one by one, they were escorted to me and afterwards allowed to leave. By the time all of the rulers had written where their stones were located in my presence, I was exhausted. My mother and Dawson watched me with concerned expressions as I took off my crown and handed it to my mother. Holding my book tightly to my chest, I announced that I was going to remain in my chambers for the rest of the evening. Hejeena kissed my forehead and agreed that I got some rest. Her affection for me as a mother was growing more and more by the hour. Dawson insisted to accompany me to my chambers.

I consented only because I needed him to tell me what the plans for my departure were. To my relief, while I had been doing the introductions, Tracy had gone into my room and left me a box containing various maps and routes of all of the different planets, as she told me she had, including Earth. I put my book on my bed and rummaged through the box, looking at all of the maps and trying to understand them. I was getting anxious. Dawson came up behind me and wrapped his arms around me, pulling me into him away from the box.

"Trish, it is time to rest. Do not overdo yourself. The members of your team know the planets. I have formed a dedicated, loyal, and hardworking team to accompany you on your journey. Allow your focus to be on finding the sword. They will guide, protect, and search for you on your travels," he said, kissing the side of my neck.

I closed my eyes for a moment, lost in the sensation of his lips on my neck. I turned to face him. He gazed at me with a look of pure adoration. I kissed his lips softly and then allowed myself to let him go. I was leaving Dawson and I was scared. Not scared in the way that I would never see him again, but more so I was afraid of losing his security. Dawson would not be coming with me on my journey, for he would be taking care of his kingdom and training a massive army of all of the best comrades from all the planets for this battle. I knew that they needed him more than me and I knew that Margaret would need him too. They would probably be married by the time I retrieved the sword, if that even happened. His father was urging the wedding to happen quickly, so that the Moon would have a queen by the time the bloody war began, just in case something happened to Dawson. The thought of Dawson dying, as an immortal, made me sick in the stomach. I would not let Bryce harm him.

Dawson watched me as I changed out of my outfit into my black evening gown. I could see the desire in his eyes, but I refused to return it. He withdrew a folder that was tucked into his shirt. He laid it on my dresser. I looked at him questionably.

"This folder contains all the information about who is coming with you on your trip. It gives you a detailed plan on navigating the space ship that you will be on, what to do in case of an emergency, how to land on certain planets, and how to radio signal the dragons. The dragons, including Draco, will spend most of their time located at different training fields scattered across the galaxy. When you radio signal the dragons, they will hear it from the radio tower located on the field they are on. There are nine different training fields within the perimeters of the planets. So you will not really have to worry about having assistance. You will be leaving in exactly two days, human time. You will be turning twenty within a few weeks and Bryce will come for you. Do not allow him to find you. You are not immortal yet and if he finds you we are all doomed."

Dawson's voice was deep and stern. He was speaking to me as if I was a child, but I knew he just cared for me. I allowed myself to go to him, one last time. I ran my fingers through his long black hair and still remembered the first time I had seen him, so long ago, at that party. Never in my life had I dreamed that a dream could be a reality. I kissed him passionately. He tightened his grip on me pulling me closer to him. His hands ran along the curvature of my waist. My sexual desire for him flared inside of me and when I opened my eyes I knew what color they were.

Dawson's eyes radiated the same energy and had turned a deep gold color. He suddenly picked me up into his arms and carried me to my bed. He then got up and placed all of the items from my bed onto the dresser and returned to lie next to me. The room was suddenly warm as he pulled me into his chest and kissed my forehead. I closed my eyes and breathed in his scent, resting my hand on his heart. His chest rose up in down, his heart beating rapidly, but I knew that he was immortal. He could not die, yet his heartbeat was still present. As if reading my mind, he put his hand on my heart and said, "Soon Trish you will become one of us."

"What is it like? To become immortal," I

asked, nuzzling his neck.

“To be honest Trish, it is like walking through the brightest light that you could ever imagine. Your immortality will come to you and it will hit you from nowhere. You will feel changed, in the lightest way anyone could be changed. You will finally be able to see and know. All of your selfish, foolish, and ignorant human traits will disappear and you will truly understand the essence of the Divine and of the Father of Eternity.”

My mind was suddenly peaceful at the thought of becoming immortal.

“Will I still be myself Dawson?”

He held me a little bit closer and traced his fingers up and down my bare arm. It gave me the chills. I looked up into his eyes, those alien blue eyes. He looked into my eyes and gave me an almost sorrowful expression.

“No, Trish you will never be the same. Your human life will fade before your eyes and your memories as a human will become vague, but you will not question the past for it is the past and you will seek nothing but the light. You will understand sacrifice, fear, and love for what it truly is."

"But Dawson I cannot allow myself to forget my past, my family, or Gabriella. I must see her again before I become immortal. What if I do not want to become immortal?" I said, stubbornly.

"Hush, Trish. Let your destiny occur in its own time. I love you."

I kissed him again. "I love you too."

I fell asleep in Dawson's arms, but it did not seem much longer until my peaceful slumber was interrupted. There was a rather odd presence in my room. I closed my eyes tightly trying to continue sleeping. I began to feel like I was having a really bad nightmare. Large smooth, claw-like hands worked their way up my body from under the covers. For some reason the touch felt erotically good. I pictured Dawson, but somehow knew that this was not his touch. These hands slowly made their way up to my abdomen, and then I felt claws slicing away my night gown. Fear suddenly took over my imagination.

"Dawson?" I tried to scream, but it came out a whimper and my eyes would not open for me.

There was no answer and I suddenly felt a heavy force pinning me to the bed. My eyes shot open, but all I could see were these pure red eyes staring at me inches from my face. A husky heavy breath blew on my face and I saw sharp glistening teeth close to my mouth. I choked back a scream. I knew it was Bryce. I felt the caress of his hot breath on my cheeks as he pulled my body into his vast cloud of darkness suffocating me with it like a blanket. I felt my body go up in flames, but they licked my skin and caressed it. I did not feel like I was burning, but my fear outweighed my power and I began to cry, uncontrollably. Why was he doing this?!

I heard him laugh evilly. "You will be mine Trish."

Then I began to see flashes of visions of humans being burned alive, chaos on the Earth, bombs, buildings on fire, children screaming for their parents. The images flashed through my mind and caused me to lose my mind. I would not allow this to happen! Every time I closed my eyes and opened them I saw nothing but pure evil.

"NO!" I screamed and my body began convulsing.

I began to scream, kick, and punch, doing whatever I could to escape this illusion that Bryce had put me under. I could still feel his claws on my body, his breath, his touch, and I wanted him gone! I knew he was enjoying this and he wanted me to share his pleasure. His laughter ringed in my ears as the lights in my chambers suddenly came on. My mother and my sisters had come into my room upon hearing my screams. I stared at their scared expressions from the door. Hejeena put her hand up to her mouth and ran to my side. I blinked, feeling dazed and weak. Bryce and his illusion disappeared. Why were they staring at me in horror? Then I looked down at the condition of my body. My nightgown had been literally sliced off of my body and there were hundreds of tiny cuts bleeding all over my skin. I began to cry as Hejeena pulled my naked body into her chest and tried to calm me down. I sobbed into her gown knowing that Bryce could only do things like this to me because I was bound to him.

"Do not let him frighten you. He cannot hurt you Trish without truly hurting himself. Do you understand me?! Now get it together! There is nothing we can do now that you are bound to the devil's son. We cannot have you feeling discouraged because of this. Madison and Rachel will help clean and bandage you. These cuts are minor and you will be healed within the next eight hours for sure. Come now. Let us get you into the bath."

As she spoke, the girls were already getting my bathwater prepared in my bathroom. Tracy went to retrieve a clean gown for me. Hejeena said she was going to make some calls, but I knew she would be calling Mavus. If Dawson found out about what Bryce was capable of doing to me, he probably would not let me leave without him. Dawson's anger was something that I did not want to see at that moment.

I allowed myself to soak in the bath water and feel the hot salt water sting the fresh cuts on my body. I welcomed the slight pain and stinging sensation, because it let me know that I could still feel anything, physically and mentally. My mind was going numb. I suddenly realized that Bryce was able to really do these things to me as if he was trying to throw me off track of my plan to defeat him. Recalling the moment when I had made the choice to bind myself to Bryce, it seemed so long ago. Then why was he, all of the sudden, just now playing these sick illusions on me? Did he know about my plan, the sword, or my departure? I hoped not and if he did who were his sources? These questions took over my mind as I sank even further into the water.

Though these answers were unclear, I knew that his power was increasing and he was trying to drift me away from attaining the sword, or otherwise drive me crazy. The nightmare he had put me through had felt so real! I finished my bath, wrapped myself in a large towel, and went back out into my room. Rachel and Madison had changed my bed sheets and prepared my bed for me to lie down on. Madison sat there and motioned for me to come lay on my stomach on the bed. Rachel was at my dresser mixing together something in a medium-sized wooden bowl. Both sisters were still in their sleeping attire and looked tired, but willing to stay by my side and take care of me.

Hejeena still had not returned. Yet, she was probably still having hysterics. I clutched the towel tightly around me and walked towards the bed. I stopped abruptly upon seeing a glance of myself in my vanity mirror. I looked like a sliced up disaster. The cuts were beginning to heal, but I looked like I had undergone someone's butcher knife. Ironically, the cuts did not really hurt, but they sure looked like they did.

"Come now, let us apply this balm to your skin," Madison said, urgently.

I looked away from myself at both of my sisters. They did not say anything else to me as I went to lay down nude on my bed. Madison allowed her eyes to look over my body, clearly distressed. She just closed her eyes for a couple of minutes and then opened them, sighing.

"What are you two putting on me?" I asked, as Rachel came over with the wooden bowl.

"It is a secret ancient recipe from mother." Rachel replied, putting a strand of her red hair behind her ear.

They both used their hands gently to massage the balm all over my body. It felt amazing. After they were finished I felt much better and relieved. Tracy had returned with a sea green gown for me to wear. To my surprise, Tracy came and sat next to me on the bed taking in my condition. A look of true genuine love came from her eyes. She began to stroke my hair. I was the happiest that I had ever been in a long time. Here we all were in that moment, Venus's daughters, the four of us, minus Sherri, who was still in a coma. But this moment was a moment of power, love, and sisterhood. It gave me strength. Looking at all of them brought tears to my eyes. Sometimes family was not necessarily rooted from where you were born or came from. Quite often the whole meaning of family extended and developed from different sources or people. I now understood that, laying there with them gathered around me tending my needs.

After I was dressed, my sisters left and returned to their quarters. There was no way I would be returning to sleep. I decided that since I was leaving in a couple of days, I ought to start packing some of my clothes. Going to my closet, I grabbed all of my jeans, t-shirts, and couple pairs of tennis shoes. I gathered my book, the maps, my necklaces, and the file of information on those coming with me. I put a bag of personal necessities together. The cellphone Griffin gave me sat on the nightstand next to my bed. I knew I would need that as well. My mother came back into my room, about an hour later.

"Meet me in my chambers for coffee. I want to show you something." She looked as if she had been crying, but her pale complexion and trembling ruby lips made her appear almost innocent and vulnerable. She just stood there staring at me. I felt for her. This had to be an emotional time for her, for all of us.

"Okay, I will meet you there. Are you okay?" I asked, looking at her hoping she would not breakdown right in front of me.

Tears were in her eyes. My goodness, I thought, I cannot deal with anymore emotions right now. If anyone should be getting all tear some over what was happening it should have been me. But a mother will be a mother no matter how unmovable she pretended to be. I began to walk towards her to comfort her, but what she said made me stop.

"I just want to know why you, of all of us, being only mortal as of now, were given this fate."

Her words caught me off guard. How was I supposed to know? If anything, I wondered the same damn thing. I did not reply, but I did not have too. She went on.

"I could not protect you long enough. You were taken from your mortal life and exposed to this reality before you turned twenty-one. If I had kept you hidden for longer you would have become immortal and seen the light sooner. Leaving your family would have been come naturally and you would be able to see your purpose more clearly, through wise eyes. But Bryce, he found you and put you through the worst disaster imaginable. Yet, look at you! You continue to not lose hope for humanity and carry on with this. This has to be the bravest choice a soul could do and try to save the universe."

I was silent for a second. I looked at her and a sense of pride came over me.

"I will not pity myself, mother. I know I will forget most of my past in a few weeks. The human life I had lived, though short, was one that I probably will not miss very much. My father, mother, sister, and brother were taken from me and I could never understand it. I changed and became coldhearted, but I still knew I had to go on. I know that they are with the Father and I know that is where I will be one day. I do not feel I have anything to lose anymore…it seems to me that my true home is not too far away," I said, my voice was strong and sure.

Hejeena smiled a small smile. "That is why…" She responded.

I raised my eyebrows. "Huh?"

She just shook her head. "I will see you in my room within the hour."

With that being said she turned to leave, but then she said over her shoulder, "By the way you have two luggage cases in your closet towards the back."

"Thank you." I said.

And that was that.

Most of everything was packed by the time I went to meet my mother. I had to admit that I was still tired due to my lack of sleep from Bryce's intrusion of my mind and body. I was going to have to obtain a carryon bag to keep the book and files close to me at all times for they contained the location of all of the stones. With the way Bryce was beginning to creep into my dreams, I had to make sure no one got their hands on that book. If I did not retrieve all the stones the universe would be hopeless.

My cuts were healing rapidly and did not look as bad as they had a couple of hours ago. Good thing I could heal quickly, I thought as I tied my hair on top of my head. The sea green gown accentuated my body nicely. I was becoming more of a woman by the day. I walked the quiet hallways to my mother's chambers. The guards lurked here and there around the castle. I did not even bother to look at them. The gloom was beginning to set in around the castle about my departure. When I walked into my mother's room she was sitting at her vanity doing her hair. She was applying some kind of treatment to her long curls. She looked at me.

"Trish,' she said, back to her usual self, 'I want to show you something. Please follow me."

She stood, her long hair flowed freely over her shoulders and back and I followed her across her room. At the far wall was a large royal blue curtain with gold tassels and embroidery that covered the entire wall. She lifted a part of the curtain to reveal a small wooden door that had an odd shaped keyhole. She looked at me, put her fingers to her lips signally silence, grabbed a candle from one of the holders on the wall and motioned for me to follow her as she unlocked the door. Without saying a word I did so. At this point, surprises did not surprise me. I figured this door led to Hejeena's towers that extended high above and looked over the kingdom of Venus.

I did not realize that ever since my abduction from Earth, I had only been in Mavus's kingdom, my mother's, and the training fields. Since I had only been inside these castle walls, I was anxious to overlook a kingdom that belonged to me. We walked up stone stairs for what seemed like forever until we reached another door. She stuck the key in the door, unlocking it. We entered into a room that was completely dark, pitch black. I froze unable to see where I was going.

"Stay right here." She commanded.

"I plan to." I replied, smartly.

I heard her footsteps hurry across the room. In the blink of an eye a blast of light over took my vision and I had to shield my eyes.

"Is it not just lovely?!" She spoke with such enthusiasm.

My eyes adjusted to their surroundings to reveal a room that consisted of an ancient looking piano, an old red vintage loveseat, several built-in wall bookshelves filled with hundreds of old books. Old papers, manuscripts, and flowers were sprawled out across the stone floor of the circular tower room. But what fascinated me the most was the balcony of which a curtain had been pulled back to reveal, that which was the source of this light. My mother stood on that balcony and it seemed that this smoky yellow haze sat over an ancient looking city. It was so bizarre because the Sun, it seemed, stood in the same place and never moved. I could not really say I felt a temperature, but I knew that I was protected. I was sure it was extremely hot out there. The buildings far away were all built of what looked to be black stone. It was amazingly beautiful, yet I saw no life.

"Where are all of the people of Venus?" I asked, looking out. The city seemed a great distance away.

She moved aside to reveal a large golden telescope.

"We that reside on Venus are not human, but

we are protected by an invisible shield. My community monitors Venus and her needs in this galaxy. The needs of my kingdom are not derived of the same needs as humans on Earth that you are used too. We are all immortal here and tied to are duty at protecting the resources and mysteries of Venus. Look through the lens to reveal your Kingdom."

Intrigued and curious, I looked through the lens. What I saw amazed me. There were these shimmering beings rummaging about in the city, but many of them had no faces or characteristics or age. They just glimmered about, these shimmering beings, taking on the human formation and bustling about the black stone city. It seemed that they interacted by touching each other because that is how they appeared to be making progress with what they were doing. All of them touched each other constantly as if it were a secret language or code. I felt a strange spiritual connection towards them and my eyes closed. A warm breeze enveloped me and I knew my eyes had turned silver.

“They feel your love and spirit and are sending you gratitude and respect as their new queen.” Hejeena said, with delight.

“This is beyond anything that I could have ever dreamed about existing.” I said, still in a sort of shock.

“Do you not know already child? To see is not to believe, to believe is to see.”

That made me smile.

“This is where I come to express my emotions on a regular basis,’ she said, motioning for me to follow her back into room, ‘At one point in time, I spent a hundred years in here in mourning. For it had been revealed to me that I was going to be giving birth to the Chosen One and what I would have to do with you.”

She closed the curtain that revealed the kingdom leaving us in pitch black darkness again. With the snap of her fingers, the entire group of black candles in the room lit up giving the room a magical dim glow. I went to sit on the loveseat taking in all of the old paintings that hung on the wall. They were of past historical figures that Hejeena had taken a liking of and there were incense lit and a hookah, old cigars, statues, and ancient artifacts and fossils I had never even seen before all over the room.

"What in the world? Where did you get all of these things?"

"Just little items I collected from my journeys to Earth over the years. You know small tidbits that the humans left behind, never found, or did not appreciate."

I was in awe at my surroundings, completely in harmony. I wrapped myself up in the classic silk sheets she had folded on the arm of the loveseat. From nowhere a small chimp came from under the loveseat. It ran straight to Hejeena and climbed up her shoulder. Startled, I sat up. Where did he come from?

"Oh, there you are Cybil. That is his name. You know he was a chimp that I discovered lost in space.

He flew into Venus's atmosphere and was bound to crash before I had him collected, trained, and he had adapted to my kingdom. He only stays in the tower, but sometimes he will accompany me on my travels to Earth, where he should have been able to enjoy his life before being used as a science project bound for failure." Hejeena said, kissing the chimp on his nose.

I went up to the small, dark brown chimp and he shyly hid behind my mother's hair.

"Relax, Cybil. Trish is your new queen and mother." She said, soothingly.

That must have helped because he bravely reached out for me to take him. I took Cybil into my arms. He wrapped his lanky arms around my neck and kissed my cheek. I could not help but smile. His big brown eyes looked at me with wonderment. "Thank you for sharing this with me." I spoke with genuine appreciation.

"Well, I wanted to save this for last, because all of this is yours. If you want you can stay your last couple of nights up here. I can have all of your stuff moved into the holding chambers before the ship arrives. You can just keep your carryon and necessity items with you. I will have them brought up here for you."

"I would love to." I responded, as Cybil held one of my fingers, moving it up and down.

"Fabulous!" Hejeena said, clasping her hands together.

"I feel that I can breathe up here and think. This solitude and peaceful environment will help me gather my thoughts and strength for this journey."

"I agree with you. Over time, my tower has helped me make many decisions in my life that helped everything go according to destiny. This is a true queen's secret quarters. I have passed it down to you and no one must enter here but you…" She spoke, boldly.

"Mother please, I beg you to allow Dawson to spend my last night with me here."

She squint her eyes and looked clearly insulted, but then she truly looked at me and saw my pain and fear. Her face softened.

"Just this once, I will send for him."

With that being said, I put Cybil down and ran over to hug her. Taken aback, she slowly embraced me.

"You truly are a child of mine."

"Did you not already know this mother?" I said, lightheartedly.

Chapter Twelve

Being a woman of her word, my mother, herself had brought up my carryon and wished me rest telling me that Dawson would be arriving shortly. I thanked her and she simply kissed my forehead and left the tower. Cybil was sitting on the loveseat eating some crackers and grapes. I had decided to look at the files that Dawson had provided for me of who was accompanying on this voyage. Turns out, there were seven people or 'beings' that were coming. Lunar and Griffin were two of them. There was a Raven, Viquino, Anson, Luca, and Tracy? Was it my sister Tracy? It could not be. Perhaps it was a different Tracy. I would find out the following day. I was at least happy that I knew and trusted a couple of them already. Lunar being Mavus's assistant and Griffin, his son.

A knock came to the door just moments after I had finished taking a peaceful rose petal bath with almond milk. My skin was blossoming with fragrance and I wore a lush white silk robe. I had my hair washed and the long curls were slowly beginning to dry. I continued drying it with the towel as I went to open the door. Cybil went and hid under the loveseat as he always did when someone came into the room. Dawson stood there in all his glory looking completely beautiful. He wore a white, buttoned collared shirt with the sleeves rolled up. It was slightly smudged with dirt. His rather high waist, black fitted pants gave him the appearance of a rider on horseback. I admired his long shiny black hair tied back revealing his mesmerizing face. By the way his dark blue eyes were gazing at me I knew he was sharing the same intimate thoughts as I was.

"Well, you look like you just finished horseback riding," I said, greeting him.

Amused he said, "Quite the contrary, I just finished riding the dragons. I was making sure each of them had memorized their routes and riding skills properly for the trip."

He glanced at the opened file on the bed. “I see you are becoming familiar early on with whom you are traveling with.”

“That is so. I am trying to get a head start.” I said, walking over to the bed to put the files away. He followed in behind me, shutting the door and taking in his surroundings.

“It is very nice up here, your mother has a rather defined taste in interior design,” Dawson said, walking around the room. Cybil came out from under the loveseat, realizing it was Dawson and ran to him hopping up and climbing onto his shoulder. Dawson chuckled petting him. I could not help but smile and watch him with a warm heart. Love was sometimes hard to explain, but its moments were precious.

“Have you ever had the opportunity to look over the kingdom of Venus?” I asked, excitedly.

He smiled at me, “No, never.”

“Well, let me show you.”

He stood there crossing his arms as I gave him

a playful smile. With a snap of my fingers, all of the candles went out and we were left standing in the darkness.

"Really, Trish," Dawson began.

I pulled the curtain back quickly already prepared to shield my eyes. Dawson quickly covered his, blinded by the vast amount of light that hit us so quickly. His eyes turned gold as they adjusted and overtook the kingdom. He walked forward in awe.

"This is just remarkable. The Kingdom where the Sun never sets," he said, admirably.

I walked up to him and he pulled me into him.

"And you are the Queen of Venus, Trish. Blaze has given much patronage for Venus."

"You are so sure of what you are saying." I said, resting my head on his shoulder.

"Why, of course! I am the King of the Moon. I am your protector and know everything," he said, his gold eyes gazing into mine. I leaned in as he bent down and kissed me.

"Dawson," I said, lustfully, "I have to have you before I leave. I may never see you again."

I pushed myself against him, bringing down his face in my hands. He closed his eyes and grabbed both of my hands, kissing them. "We cannot make love Trish until you become immortal and even then it is probably not a good idea. I had a vision, not too long ago that you were impregnated with my son and to make matters worse, I lost you both."

Tears formed in my eyes and I looked away. "No, Dawson that would never happen."

He shook his head. "I will not risk it. If we all survive through this, then I would want nothing more than to make love to you, call off my engagement and have a family with you. However, I cannot live in hope right now simply because I know that circumstances may very well go otherwise. If my vision came true and I lost you and my child I would kill myself. I cannot even imagine having a half-immortal and half-mortal child right now, while you are bound to Bryce."

That stung my heart and I felt a tear slip down my cheek. "Dawson you speak wrongly! Your vision is just your fear playing mind tricks on you. I thought you saw the light Dawson!" I said.

His eyes were such a ravenous gold from the blazing sun. "Look into my eyes Trish. Look into them," He said, forcibly.

Reluctantly, I looked into them. What I saw in his eyes made me squirm from his grip on me. They were a hungry orange with his black hair tied back revealing his entire face he looked like he wanted to devour me, not literally but in desirable sense. There was a sudden golden glow that radiated from my black stone kingdom and Dawson again pulled me back into his embrace.

"Your kingdom desires for us to be as well. You connect the moon to the sun, Trish. You are the connection of the light and of the dark. I love you, remember that, but part of seeing the light Trish, came with knowing when to sacrifice and not give into temptation even at the hardest moments of life. This is by far the hardest moment in my immortality since I can remember, letting you leave. I still remember you, picking you up from that party. I thought, darn it father, why did she have to be so innocent, so beautiful?"

I blushed. "No, you did not."

We moved away from the balcony back into the room leaving the curtain open. It gave the room that sunset glow, but it was permanent and peaceful. Not too bright and not too dark.

I grabbed Dawson's hands, "Stay with me until I depart tomorrow."

He smiled at me, "Silly miss, I planned to, but do tell me where will we rest?"

That was a good idea. As I began to think Cybil scurried from where he had been lounging on the balcony and went to a wall in the room that had an old Asian themed tapestry on the wall. Cybil climbed up on the tapestry where there was a slight stone sticking out from the wall. Dawson and I watched, clearly impressed by Cybil's talents. The chimp pushed the stone into the wall and the whole length of the tapestry wall came down slowly revealing a massively large bed. No wonder the room was quite bare and scattered with papers. The balcony was visible from the bed and it looked like pure sleeping heaven. It was what I would call a California-King circle bed, because that is exactly what it was covered with huge royal red velvet blankets, white silk sheets, and huge cotton pillows and when we went to lay down in it, it seemed to absorb us into it.

"This is great," I said, dreamily as I allowed myself to sink into the mattress.

Dawson went to unbutton his shirt and I decided to resist watching him and look back out at the kingdom. Cybil showed me where there was a private smaller balcony on the other side of the tower. It had a handmade wooden shower head hanging from a long plant vine that extended along the side of the tower.

Hejeena and her ways, I thought. But amazingly enough, when one pulled the rope next to it light clean green water came out of the faucet.

"Well, we can take a shower when we wake up." I said, coming back into my room.

Dawson lay there on my bed watching me discover all the little secrets that Cybil was showing me. I had giggled as Cybil showed me where Hejeena kept images of all of the famous men of Earth she had found attractive. Dawson smiled at me and never took his eyes off of me, but he looked so peaceful and relaxed. I had given him the file because he wanted to double check that I had everything for my trip. I knew he was here to keep me calm in these last twenty-four hours before I left and I am glad he did so because, truthfully, I was terribly frightened. Cybil was doing a great job of keeping me distracted, but I knew the time was coming. I just sat there, on the balcony ledge staring out at my kingdom and looking out of the gold telescope. Dawson grabbed my attention.

"Have you read through all of these files and become familiar with who is coming with you on your quest?" He said.

"Yes," I said.

He gave me a look.

"Well, I know I do not know everyone that is coming with me, like personally, but Lunar and Griffin, I know them."

"Tracy?" Dawson said.

Tracy, oh yeah! I had forgotten.

"Yes, is that my sister or a different Tracy?" I asked.

"Come on Trish, be practical. It is your sister who had decided to add her name to the docket."

I was so happy. Dawson gave me another one of his looks. "I worry for you, my love."

Shoot, I worried for myself too. I went and retrieved my book from the bed. Dawson went back to looking at the files. "I am confident everything is in order and that you will be with the ones I would trust with my life on your trip."

"I believe I will be alright." I said, reassuring myself and him.

I opened my book to see where some of the rulers had written the location of their stones. My heart froze as I read what Blaze wrote on one of the pages. His cursive was angrily scribbled across the page, but his demeanor had been so calm. What I read really ignited a flame somewhere from within in my body. It read:

Trish, look at the Sun. Really look at it, for you are able to do that out here in these worlds, but on Earth they cannot see the Sun for what it really is. My kingdom has always been the home of where the souls of the damned and the truly unredeemed went. Once the Sun receives a soul it is there for eternity as promised by the Father. The sun is the Lake of Fire, the Star that continues to shine over the earth to provide guidance and hope over the lost world. The Sun gives it light. Never had I had to discuss my birthstone location with any soul in existence, but you have to do this and I must tell you. Denmark, Germany... that is where it is. Within the grave of B.T. Good luck Trish.

I felt hot all of the sudden and looked at the Sun, truly looked at it over the horizon. I closed my book tears in my eyes staring at the sun. It was too much to comprehend, but I knew that I had too. It was the Sun's, Venus's, and the Moon's dark connection that somehow the Earth depended on. I knew where I had to go first and I finally was beginning to know who I was. After putting my book away and coming to realization, almost as if I was in a trance went to lie next to Dawson. He had been watching me, as he always did, without saying anything. He took me into his arms as I began to cry at the reality of what was truly going on.

"Why did you never tell me what the Sun was?" I asked, weeping.

"You had to learn for yourself, Trish. As the Chosen One, you have to experience your own transition to become who you are meant to be. The Sun is not any different because of what you know. It is still the daily source of light for the humans and provides warmth and comfort to humanity, but it has a deeper hidden purpose as well. It just contains the fire of eternity, that is all, Trish."

I sobbed into his chest until I could not cry any longer. He kept holding me, stroking my tangle of curls. "I guess this could be expected, you are after all twenty tomorrow."

"What?" I gasped. "But that means I

will…Bryce will be coming for me! Dawson how will I escape him? What is going on, do not let this happen?! Time has gone by too fast." I began to start freaking out.

"Trish, calm down. I will protect you from him, but you will undergo a slight transition tonight because of your birthday. I am staying with you tonight during the lunar eclipse that will take place the moment you turn twenty. There is a full moon tonight, so all of my brothers and sisters are on their toes and navigating the galaxy making sure everything is fine. There is a lot of power across the universe tonight as a result of your birthday transition. Your power will be unleashed in immeasurable ways as you get closer to your immortality. Now you will begin to understand that there is no such occurrence as time. We are forever under the Father. Just rest here and when you wake up you will be slightly different. I will be right here through the whole thing and will make sure that you do not even notice it take place." Dawson said.

My eyes had turned silver and I was shivering, uncontrollably. Dawson pulled me closer to him and kissed me several times. But I had gone into shock. Everything was driving me insane and I think I was going to have my breaking point. Dawson tried to get me to lie down, but I fought him off.

"No, I want to see this."

"Trish, I think it is best you should rest."

He got up to close the curtain, but I screamed, "Leave it opened, Dawson! I want to see it for myself!"

His jaw clenched, but he left it open and walked to the far side of the room touching his forehead. I just stared at the sun with my red tear brimmed eyes. He came to me pulling me into his arms.

"My brother has called and informed me that the lunar eclipse will be taking place in the next hour. I have ordered for your ship to be prepared for departure. Your team will be arriving just after the eclipse. They will meet you aboard the ship. Lunar is going to be navigating the ship. Raven, Viquino, and Tracy will be accompanying you off the ship wherever you may need to go and Anson and Luca will stay on the ship to monitor any nearby evil power or energy. They will keep a look out while communicating with the rulers of the planets. All is underway, my love. Since, I can see you are rejecting sleep, I will assume that you will take rest on the ship for your travels may take several days, but the ship is large enough that everyone has their own rooms, there is an office, the pilot and control booth, a kitchen, and training area. It is the best ship we have created yet."

Dawson was such a warrior at heart. Determined and informative, he wanted me to be on top of everything. I looked at him with all of the love I had in my heart. He suddenly looked a little afraid and moved away from the bed. I found myself moving towards him.

"Trish, do not do this right now."

I stopped. “I am sorry, Dawson. I just love you so much.”

“As I love you, but control your power, before it controls you. There is too much of it surging through the universe right now that it is suffocating.”

Suddenly, something changed and caught my attention. The room was slowly growing darker.

“What is happening?”

“I told you the lunar eclipse. Your transition will occur shortly and you will release a great amount of power from within yourself without being able to stop it. Just embrace it and let it go, but you may feel completely drained afterward. It is a transitioning moment, a very powerful moment, that is why I had hoped you would be sleeping, but since you are awake I will try to help you not drown in this power.” He came to me, shirtless, and my skin crawled with excitement.

Closing his eyes he pulled me up off of the bed into his arms and carried me to the balcony. Just his touch was enough to send volts of my power out. It caressed both of us and the warm wind I felt coming from my kingdom now hit us both with double the force. I was no longer scared, but the freest and happiest I had ever felt in my life. The moon was suddenly beginning to pass by the Earth's shadow, but somehow as a result there was a dim overcast set over the Sun. I looked at Dawson and he looked at me. His gold eyes were vivid in the darkness and he held me close.

"Are you ready for this Trish?" He asked.

"I suppose so," I said.

"Good," he said, and continued to hold me close in the darkness.

Something was wrong though.

"If I am bound to Bryce, will he be able to hurt me during this transition? His ties into my mind will weaken right? I will be more powerful now that I am twenty and even closer to immortality. And, he will have even more difficulty finding me, right?"

I felt him tense from the words I spoke.

"That is correct Trish." Dawson said. "But please do not speak of him. There is too much power in the air and you are so close to your transition."

But it was too late. All I had to do at this point, I had not realized, was speak of him. He had already broken into the protected shields of Venus's atmosphere and was heading towards the kingdom. All of this somehow flashed before Dawson's and my mind, as if we received some sort of telepathic emergency message. Dawson, wasting no time, touched his forehead most likely giving orders.

"Do you trust me?" He said, looking into my eyes dead serious.

"Yes, Dawson, what is going on?" I was frightened.

Suddenly the darkness seemed darker and there was a horrible, deadly distant howling in the wind now.

"Go grab your bag and run back down the

tower. Take the route to the training fields where the dragons are. There your ship will be awaiting you with everyone on it. There will be chaos Trish, but do not stop, no matter what you see. Swear to me that you will do this! There is no time to waste! I hope that you will make it to the ship so that your transition will take place before the eclipse is over. You have five minutes to get there, go, now. I love you, but I must go hold him back now."

Dawson kissed me roughly, but passionately. I began to cry and scream as I ran to retrieve my bag and rush out of the room. Words could not explain my grief as I refused to look behind my shoulder. I was running in complete darkness. The wails and howls seemed to get louder and closer as I ran and I heard a lot of commotion and rumbling. When I emerged from the tower, I saw the guards running about yelling and putting barricades on the castle entrances. I heard shrieks, screams, and horrifying roars outside of the castle walls as I ran even further towards the training fields. I began to have all sorts of hot flashes and visions of Henry, Nikki, Gabriella, my father and my mother. I could not make out where I was going, but I continued to run. It felt like unforeseen hands were grabbing at my skin and pulling at my hair. I began to feel those slices again ripping against my skin, reopening old wounds. I cried out in pain.

I prayed Dawson was alright, because I felt as if I were dying. The double doors to the field came into view and I rushed out them into the open field. The sky was scorching red and I saw Draco among all the other dragons fighting what seemed to be huge black dragons of smoke with horrifying red eyes. About half a mile across the field, was a massively large, jet blue space ship. The entrance of the ship was lowered for me to hop up onto. It seemed so far away, but I could make out the distant figures of my team waiting for me, calling out to me.

“Trish!” I heard from behind me, above all of the wails and roars of the fighting dragons.

Tears continued to fall as I looked behind me. The wind was so fierce that debris began to blow violently around the field. The wind whipped harshly against my skin making me cry out in pain even more. It was my mother that I made out standing at the doors that I had run out from.

"Trish, run! Hurry, for he is coming!" She yelled.

I began to run as fast as I could towards the ship, but the wind seemed to blow against me harder the closer I got to the ship. I screamed, furiously as I pushed myself against Bryce's power. It cut me, I felt the blood seeping through my gown and dripping from my skin. Still I continued on. Venus's kiss burned too hot from underneath my gown and I could see it glowing bright orange as if it were on fire. It began to burn a hole through my gown and I clutched my stomach. It would not be able to protect me this time. I looked up and saw that the Sun was slowly revealing itself.

This was it, I thought to myself. A couple of the dragons fell dead from the sky, my tears turned to blood, and I saw Bryce emerge from the dark stormy clouds. He stood there, his black eyes locked onto mine and I froze only a few yards from the ship. He was ready for me, I could tell. The team had run onboard the ship. Tracy was the only one who stood there reaching and yelling out to me her arms open. But I could not move. Bryce smiled at me as he continued towards me, on his black cloud. It carried him naturally across the sky. With all of the power I had I threw my carryon bag to Tracy, she caught it effortlessly.

I knew what was coming. Bryce was going to try and capture me and take me to the underworld before my transition. I would not allow this to happen. It could all end now. I turned and began to run the other way. The wind carried me faster the other way, my endurance increased, and I ran further away from Bryce. This angered him. I kept my eyes on the Sun as it slowly began to emerge from the overcast of darkness. Bryce picked up speed and trailed me closely. There is no such occurrence as time, I reminded myself as I ran. It seemed as if I had wings carrying me as I ran for I increased my pace with ease.

"Give into me!" I heard his voice, as if he were whispering into my ear.

My fear grew, but so did my power. I ran faster and faster as the Sun came more into view.

"Stop at once or I will kill you!" Bryce screamed.

"No!" I screamed, losing hope, but continuing to run.

He was literally right behind me now. I felt my heart grow heavy as I looked at the Sun. My strength was weakening and I felt like I would fall, but I kept running. I looked behind me and Bryce had his arms reaching out for me. His black eyes, black hair, sharp teeth, skin the color of death, all of him was closing in on me. The Sun was almost fully revealed, and I felt the warmth of the light peak from the clouds. Bryce began chanting something and I felt like I was being sucked back towards him.

"No! Father, help me!" I screamed, reaching out towards the Sun.

As if answering my prayers, I looked behind me just in time to see Dawson take out Bryce. He had come from nowhere and suddenly Bryce's control over me was gone. He and Dawson were wrapped in a billowing cloud of darkness. At that moment the Sun's light fully blasted from the darkness that had engulfed it and then it happened. My transition took place. The rays of the Sun had never felt so good. I heard Bryce's screams of pain as he disappeared, the black dragons disappeared, and I flew up into the air. All I could see was light and all I could feel was light. I cannot even begin to explain what I felt. All I could say was that I was finally whole and cleansed. There was nothing that I did not know or did not understand, everything made sense and my heart beat the fastest it had ever beat. It was my soul that was once void, that was filled with the light. To see nothing but true light and wisdom was something that never really came until the end of one's natural life in the mortal world, but as my transition took place, the universe and its purpose and mine came to materialize in my mind.

Now, I no longer felt pain, fear, or anything negative. I knew no emotions except love, duty, honor, loyalty, dignity, and strength. My past was gone, but I still remembered those people in my mortal life that had meant the most to me. Now I understood their true purpose. Death and life were reciprocals of each other that I thoroughly understood as well. The light inside of me stored the darkness that was once equally present inside of me away so that it had no hold over me anymore.

Finally, I felt free from the chains that held me and I knew that though my journey was just beginning, I had the willpower to do it now. Bryce was gone for the moment, his ties to me broken and the deal he and I made might as well have never taken place. He would have much difficulty finding me now as I searched for the stones and for the sword. I did not know what would happen when I awoke from the light and my transition, but I knew that my journey towards ultimately saving the universe would continue on. I had only survived the first round and the war between good and evil had yet to begin.

About the Author

The Stars Touch Here is Amilee Turner's second published novel and the first part of a sequel. She currently resides in St. Louis, Missouri. Miss Turner is a self-published author, model, actress, and a current law student.

Made in United States
North Haven, CT
04 November 2026

21765353R00203